GENERAL MATHEMATICS
IN
AMERICAN COLLEGES

By KENNETH E. BROWN, Ph.D.

TEACHERS COLLEGE, COLUMBIA UNIVERSITY
CONTRIBUTIONS TO EDUCATION, NO. 893

*Published with the Approval of
Professor William D. Reeve, Sponsor*

BUREAU OF PUBLICATIONS

TEACHERS COLLEGE, COLUMBIA UNIVERSITY

NEW YORK · 1943

ACKNOWLEDGMENTS

Most dissertations are the result of the cooperative effort of many persons, with the burden falling heavily upon the sponsor; and this investigation has been no exception. The author is deeply indebted to his sponsor, Professor William D. Reeve, for his continual guidance and constructive criticism.

In addition, the author wishes to express his gratitude to Professor John R. Clark, Professor Clifford B. Upton, and the members of the Mathematics Seminar of Teachers College, Columbia University, for many valuable suggestions; and to Miss Margaret Berger for encouragement and suggestions in collecting the data and in the preparation of the manuscript.

The author also gratefully acknowledges his obligation to the many teachers and students who have contributed to this investigation by so promptly returning questionnaires and answering correspondence. Special acknowledgment is due the teachers and students of the following institutions whose cheerful cooperation has helped to make the study possible: Barnard College, College of the City of New York, Cooper Union Institute, Brooklyn College, Hunter College (Manhattan and The Bronx), Mills School, Manhattanville College of the Sacred Heart, New Jersey State Teachers College (Jersey City, Montclair, and Paterson), Queens College, Swarthmore College, and the University of Newark.

To his wife, Kristine Brown, the author is especially grateful for unfailing confidence and cooperation during the preparation of this study.

K. E. B.

New Jersey State Teachers College
Paterson, New Jersey

CONTENTS

Classroom Observations. Summary of Provisions for
Meeting the Objectives of General Mathematics.

Success as Indicated by Authors of Preparatory Textbooks. Success as Indicated by Authors of Cultural-
Preparatory Textbooks. Success as Indicated by Authors of Cultural Textbooks. Success as Indicated by
the Questionnaire. Success as Indicated by Classroom
Visitations. Success as Indicated by the Students' Questionnaire. Summary—Success in Meeting the Objectives
of General Mathematics.

Purpose of Study. Reports of Other Investigations.
Rise of General Mathematics Movement. Objectives of
General Mathematics. Provision for Meeting the Objectives of General Mathematics. Success in Meeting
the Objectives of General Mathematics. Crystal Gazing.

GENERAL MATHEMATICS

IN

AMERICAN COLLEGES

CHAPTER I

INTRODUCTION

THIS study had its origin in a quest for suitable material to meet the mathematical needs of seventy young girls at Mills School for Kindergarten–Primary Teachers, New York City, who were attempting to meet the certification requirements of New York for kindergarten and primary teaching. What are the objectives that other teachers have proposed for non-major mathematics students? What type of material are other instructors offering in such courses? Do they give a review of arithmetic and high school mathematics? Do they just talk in general about mathematics? What success are these teachers having in attempting to meet life's mathematical needs of the large academic group of students who seek other fields of study after the freshman year?

The desire to find answers to these and kindred questions motivated this investigation.

PREVIEW

In harmony with this purpose a review of the related pertinent studies is given in Chapter II to serve as an orientation to previous research in this area. A bibliography is included in the Appendix for the convenience of those who may wish to acquaint themselves more fully with the work of others in this field.

To give a perspective and serve as a frame of reference from which to examine the freshman general mathematics course, the rise of the general mathematics movement is described in Chapter III. This examination of the literature in the field was made, not to provide an exhaustive treatise on the history of general mathematics but to serve as a setting that will prevent a distorted view of the major problems of this dissertation.

1

Chapter IV contains the discussion of the objectives of general education, with special reference to the aims and purposes of general mathematics. The objectives of general mathematics are those that have been indicated by (*a*) reports of committees of specialists in the field, (*b*) authors of the textbooks, and (*c*) teachers of the subject, both by their statements and by their demonstrations in the classroom.

Provisions for meeting these objectives are presented in Chapter V. The content of the courses, the style of presentation, and the emphasis given to topics are presented not only as indicated by the authors of the textbooks and the teachers of the subjects, but also as observed in the classroom.

Success in meeting these objectives of general mathematics is the general theme of Chapter VI. The data of this chapter are based upon recorded classroom observations and the opinions of authors, instructors, and students.

Chapter VII concludes the report of this study with an evaluation of the entire investigation.

REITERATION OF PURPOSE

Briefly stated, the aims of this study are:

1. To trace the historical development of college general mathematics in the United States.

2. To show the present status of general mathematics in American colleges.

3. To discover and point out certain trends in the development of college general mathematics.

THE DATA

The data used in this study are based upon: (1) A survey of the pertinent literature in the field. (2) A questionnaire answered by 458 colleges in the United States offering general mathematics. (3) An analysis of more than fifty general mathematics textbooks. (4) Recorded observations of fifty general mathematics classroom recitations. (5) Opinions of 1,500 students enrolled in general mathematics classes.

1. In harmony with the purpose of this study, the survey of

the literature laid primary emphasis upon the history of the general mathematics movement and the objectives of general mathematics on the collegiate level. Although it was not thought advisable to present a detailed analysis of the relevant literature, applicable excerpts have been included throughout the study.

2. Preliminary to the construction of the questionnaire, an examination was made of the catalogues of 1,266 universities, colleges, and junior colleges. The list of colleges was secured from the 1940 Educational Directory of the United States Office of Education. The catalogues that were not available in the Teachers College Library, Columbia University, were secured by writing directly to the colleges. The description and content of the mathematics courses offered by these institutions were carefully examined. If the description or title of a course in any way indicated that it might be general mathematics, the head of the department in the institution offering such a course was sent Questionnaire A.[1] A letter of explanation[2] accompanied the questionnaire. Of the 1,266 schools represented by the catalogues examined, 492 were sent the questionnaire. To the institutions which did not return the questionnaire within a reasonable time, a second request[3] was sent. As a result, 458 replies were received from the 492 schools.

3. In securing the titles of the general mathematics textbooks to be analyzed, the investigator made use of Reader's Guide to Periodical Literature, Publishers Trade List Annual, Cumulative Book Index, Educational Index, the card indexes in the libraries of New York University, Teachers College, and Columbia University, and the responses to Questionnaire A. More than fifty textbooks whose titles were thus secured were analyzed according to style of presentation and percentage of pages and exercises devoted to various topics in mathematics. The topics from the textbooks fell into approximately sixty categories. For convenience of comparison these topics were grouped in approximately twenty-five divisions similar to those

[1] See Appendix B.
[2] See Appendix A.
[3] See Appendix C.

recommended in the report of the Joint Commission of the Mathematical Association of America and the National Council of Teachers of Mathematics.[4]

4. The recorded observations were the result of more than fifty visits to the classes of thirty-seven college teachers of general mathematics in twelve eastern colleges.[5]

The opinions of the students and instructors were secured from conversations during the visits, and in particular the reactions of the students to the course were secured by the use of Questionnaire B.[6] This questionnaire was given to more than fifteen hundred students. The fact that many questionnaires were not only filled out completely but contained numerous added suggestions is indicative of the student interest in the subject. The interest of the instructors was indicated by their hearty cooperation in giving during the class period all the time necessary for filling in the questionnaire.

Although a complete record of the detailed analysis of textbooks, class observations, interviews, and questionnaires is neither feasible nor desirable in this report, sufficient material will be presented from the authors of textbooks, instructors, and students to give a clear picture. An attempt has been made not to offend either by overcondensing pertinent material or by obscuring the findings by placing them in unwieldy tables.

In regard to the next chapter, perhaps if it occurred in a certain type of general mathematics textbook it would have the following note: "Competent students may omit this chapter but it is the author's belief that even for them it will be valuable as a review."

---

[4] The Final Report of the Joint Commission of The Mathematical Association of America and the National Council of Teachers of Mathematics, *The Place of Mathematics in Secondary Education*, pp. 159-161. Bureau of Publications, Teachers College, Columbia University, New York, 1940.

[5] See Acknowledgments.

[6] See Appendix D.

CHAPTER II

REPORTS OF PREVIOUS INVESTIGATORS

DEVELOPMENT OF COURSES OF STUDY IN GENERAL MATHEMATICS

AS early as 1915 there were discussions of attempts at formulating courses in general mathematics. A report of one of these attempts is given in the *American Mathematical Monthly* by F. L. Griffin, who describes the course at Reed College.[1] The course at that time was divided into the following topics: (1) some practical uses of graphs; (2) some important limit concepts; (3) differentiation; (4) integration; (5) trigonometric functions of acute angles; (6) logarithms; (7) further differentiation and integration; (8) uses of rectangular coordinates; (9) solution of equations; (10) polar coordinates and trigonometric functions in general; (11) trigonometric analysis; (12) definite integrals; (13) progressions and series; and (14) probability and least squares. A quarter of a century later Professor F. G. Reynolds, head of the department of mathematics at City College, New York City, stated in an interview with the writer that this radical departure from the traditional accepted courses for freshmen was one of the outstanding contributions to mathematical education during the period 1900–1920. The material of this course was later published and became a popular textbook.[2]

It is not to be assumed that this was the first textbook published in the field. Rather, by 1915, as has been pointed out by others,[3] general mathematics had reached a definite stage in evo-

---

[1] F. L. Griffin, "An Experiment in Correlating Freshman Mathematics." *American Mathematical Monthly*, Vol. XXII, pp. 325-330, December, 1915.

[2] F. L. Griffin, *An Introduction to Mathematical Analysis.* Houghton Mifflin Company, Boston, 1921.

[3] J. W. Young, "Organization of College Courses in Mathematics for Freshmen." *American Mathematical Monthly*, pp. 6-14, January, 1923.

lution, and articles began to appear regarding the experimentation that had been taking place in the previous years.

As partial fulfillment for the doctorate in 1932, Wright[4] reports the formulation of an outline for a course in general mathematics designed to meet the needs of students preparing for the study of the calculus. The criterion used in selecting the material for this outline was based primarily upon the mathematics needed in the solution of the exercises in Ford's calculus textbook. Wright also studied the catalogues of 205 colleges and universities with regard to the trend in requiring general mathematics as a prerequisite for the study of the calculus. He reports that over fifty of these 205 schools offered a course in general mathematics, and in over one-half of them it was preparatory to the calculus.

In 1935 Trainor[5] reported briefly on the course in general mathematics which was being developed at Washington State Normal School. The departure of this course from the usual traditional freshman mathematics course may be seen in both content and objectives. Some of the topics included were: the nature of mathematics, the significance of axioms and postulates, invariance, the theory of groups, higher dimensionality, imaginaries and transcendentals, infinity, non-Euclidean geometries, and topics concerning the history of the subject. Trainor states the aims of the course as follows:

1. The demonstration to the students that mathematics is a "live" subject today.

2. The pointing out that the most significant human thought which the race has is in its higher mathematics.

3. To show that the processes of thought used in higher mathematics have significance for everyday thinking.

4. To have the student become acquainted with some very important recent literature on the subject.[6]

---

[4] H. A. Wright, "An Evaluation of Certain Textbooks in General Mathematics for College Freshmen with a View to Formulating a Course Which Affords More Satisfactory Preparation for Calculus." Unpublished Doctor's dissertation, New York University, 1932.

[5] J. C. Trainor, "A New Approach for a Course in Mathematics for Teachers." *School and Society*, 41:398, March, 1935.

[6] *Ibid.*

Georges[7] reports the formulation of a course in general mathematics based not upon the objectives stated by Trainor, but upon the propaedeutic requirements of students. The determination of the needs of the students was based upon an analysis of textbooks in the fields of biological science, chemistry, economics, geography, geology, physics, psychology, and sociology, and conferences with instructors in these subjects concerning the mathematical concepts, principles, and processes that can be used advantageously. After enumerating the concepts thus determined, Georges states: "That these findings are but a restatement and reiteration of the fundamental, basic, and pillar concepts of mathematics needs no comment." [8]

In order to determine more adequately the desirable mathematical preparation of students, Richtmyer[9] secured the cooperation of 389 teachers and administrators in checking a list of eighty-two mathematical items as to frequency, importance of use, and difficulty in learning. On the basis of this information regarding the mathematical needs of prospective teachers, a general mathematics course was organized and published as a textbook.

Content of General Mathematics

During the development of courses in general mathematics there have been certain attempts to analyze the content of these courses. Although the number of investigations have been few, the following examples indicate both the type of research and the effort made to develop the content material.

In 1923 Sanford[10] reported an analysis of eleven mathematics textbooks for college freshmen. The textbooks were published during the period 1910–1920. Sanford compared the topics

---

[7] J. S. Georges, "Mathematics in the Junior College." *School Science and Mathematics,* 37:302-316, March, 1937.

[8] *Ibid.,* p. 312.

[9] C. C. Richtmyer, "Functional Mathematics Needs of Teachers." *Journal of Experimental Education,* 6:396-398, June, 1938. Also C. C. Richtmyer and J. W. Foust, *First Year College Mathematics.* F. S. Crofts and Company, New York, 1942.

[10] Vera Sanford, "Textbooks in Unified Mathematics for College Freshmen." *The Mathematics Teacher,* 16:206-214, April, 1923.

listed in the report "The Reorganization of Mathematics in Secondary Education," [11] and concluded there was not only little agreement between the textbooks and the report, but also little agreement among the textbooks as to the important topics. She pointed out that many of the textbooks contained little unification, but that they did indicate that an attempt was being made to unify the separate branches of mathematics for freshmen. Sanford predicted that "the future development of this work would be less and less along the conventional lines."

Anderson[12] reported that an analysis of the catalogues of seventy colleges and universities indicated that the general mathematics courses were organized around such central ideas as a fusion of topics from traditional mathematics, historical development of the subject, or some main theme such as statistics, functions, or mensuration.

In an analysis of the general mathematics offered at the Iowa State Teachers College, Watson[13] reports that "A student would have occasion to use, either directly or indirectly (*a*) 350 of the semi-technical words of plane geometry, (*b*) 168 of the basic ideas, (*c*) 65 of the essential theorems, (*d*) 15 theorems which are not among the essentials, (*e*) 1,400 new semi-technical words." The conclusion from the data furnished by the analysis was that a "fair knowledge" of plane geometry and "basic ideas" of high school mathematics is necessary for success in this general mathematics course.

VALUE OF GENERAL MATHEMATICS

While leaders in the field of general mathematics education have been attempting to formulate a course which would meet the needs of the "non-specializing" academic group, critics of mathematics have been questioning the mathematical training

[11] A Report by The National Committee on Mathematical Requirements under the auspices of the Mathematical Association of America, *The Reorganization of Mathematics in Secondary Education*, pp. 37, 38. U. S. Bureau of Education Bulletin No. 32. 1923. Washington, D. C., 1923.

[12] Frank Anderson, "The General Mathematics Course in Higher Institutions," pp. 11-12. Unpublished Master's thesis, University of Arizona, 1938.

[13] E. E. Watson, "An Analysis of Freshman College Mathematics." *Education*, 48:225-228, December, 1927.

given by the general course. The advocates of general mathematics courses have cited experiments which they believe indicate that general mathematics is a better preparation for students in high school[14] and they have suggested that it is also better preparation for those in college.

On the other hand, critics have contended that the mathematics courses from which the materials from several branches of mathematics are drawn are not well organized; and if these general courses were well organized, they would be confusing to the student. Also the critics have maintained that it is only when the student has a thorough grounding in the individual subjects as they are studied separately that he will be able to correlate the material. Some of the critics have contended that it would be possible for the student to study the subjects fused around some unifying principle after he has a sufficient knowledge of the individual subject, but not before. Others feel that the study of the individual subject is sufficient and that the material will automatically be integrated by the student. Still others contend that the material is just not well learned unless it is studied separately.

Leaders in favor of general mathematics have criticized the compartmental plan of teaching, calling attention to the fact that life does not bring our problems to us in air-tight compartments. They maintain that it is much better for the student to understand a mathematical subject with its relationships to the entire field of mathematics. They have called attention to the valuable research papers and the success in the advanced courses by students of general mathematics. In contrast, the large number of failures now in mathematics and the lack of an understanding of mathematics by those who do pass the course are pointed out by the critics. So the debate has gone on for several years.

Although workers[15] had experimented with methods of teach-

---

[14] C. McCormick, *Teaching of General Mathematics in the Secondary Schools of the United States*, pp. 32-53. Bureau of Publications, Teachers College, Columbia University, New York, 1929.

[15] R. R. Davis and H. R. Douglas, "The Relative Effectiveness of Lecture Recitation and Supervised-Individual Methods in the Teaching of Unified Mathe-

ing general mathematics, no one had reported the evaluation of the relative efficiency of the mathematical preparation given by the general mathematics courses compared with that of the traditional courses. It was in answer to these critics of general mathematics that Scott[16] in 1939 reported an experiment conducted under his supervision. In Scott's study, the relative achievement in mathematics of two groups of college freshmen was compared. The control group studied the traditional courses in freshman mathematics, whereas the experimental group studied the general mathematics course.

Scott's study comprised ninety-five students in each group who were taking freshman mathematics during the academic year 1937–1938. The control group was composed of students enrolled in the Louisiana State University; the experimental group represented students from George Peabody College for Teachers, College of Mines and Metallurgy (University of Texas), Texas Christian University, University of Chattanooga, and Louisiana State University.

The comparison was made by pairing students on the basis of previous achievement and according to the number of hours devoted to recitation during the session. This procedure necessitated the dividing of both the experimental and the control students into three groups. Comparisons were made between two groups of forty-seven students who took six semester hours of mathematics, between two groups of forty-eight students who took nine semester hours of mathematics, and between all the experimental students and all the control students.

The students in the experiment represented eight different institutions and they were taught by fifteen different instructors. They were given an achievement test at the beginning of the session and near the end of the second semester. These tests were designed to measure achievement in algebra; algebra

matics in College." *Controlled Experimentation in the Study of Methods of College Teaching*, Vol. 1, No. 7, p. 313. University of Oregon Publication, Eugene, Oregon, 1929.

[16] P. C. Scott, "An Abstract of a Comparative Study of Achievement in College Freshman Mathematics." George Peabody College for Teachers, Nashville, Tennessee, 1939.

and trigonometry; and algebra, trigonometry, and geometry. The difference in achievement was greatest in the comparison of the experimental group of students with those students of the control group who would probably take only one year of mathematics. However, the difference in the relative achievement of the engineering students and the general mathematics group was not so marked, but in all the comparisons of the study, the achievement of the experimental group was reported to have been higher.

Attention should be called to the fact that although the gains were slight as measured by the tests, the tests examined for only certain facts and processes which may not be the entire or even an outstanding objective of a general mathematics course. Perhaps the results of these scores do not indicate the "relative achievement" of the two groups. Scott no doubt had this in mind when in the "Suggested Subjects for Further Study," he states: "The aims of the recent college freshman textbooks in general mathematics go beyond that of teaching a mere knowledge of the facts and processes of the fused subjects. These aims include teaching and appreciation of mathematics as a cultural subject, showing the wide application of the science to many of the fields of knowledge, and providing a standard of exact reasoning. To investigate the extent to which general mathematics courses fulfill these aims would be a valuable study."

Scott has not attempted to answer this question but he has contributed to this particular field by trying to answer the question, "Do students in general mathematics develop as thorough an understanding of algebra and trigonometry, and as much skill in using the techniques of algebra and trigonometry as those students enrolled in the traditional freshman mathematics courses?"

Popularity of General Mathematics

Several workers in the field of mathematical education have attempted to investigate the response of the teachers to the courses in general mathematics. In fact, an entire session of the Seventh Summer Meeting of the Mathematical Association of America was given to the "Present Status of Unified Mathematics." It was at this meeting that Young[17] reported the results of information received from ninety-eight institutions. Of these institutions fifty-nine stated that they had given a course in unified mathematics. Forty-one also indicated the years in which unified mathematics was offered and from this information Young concluded that the number of colleges offering general mathematics in the following years were: 1917–18, 14; 1918–19, 20; 1919–20, 27; 1920–21, 23; 1921–22, 23.

It is pointed out in the report that because of incomplete data too much reliance should not be placed on these figures in establishing a trend, but it probably could be safely concluded that a "strong body of institutions" are in favor of the course.

In 1929 Hills[18] reported that a survey of junior college mathematics based on a questionnaire showed that thirty-six of the reporting eighty-eight junior colleges offered a unified course in mathematics. He also states that approximately 20 per cent of these colleges were dissatisfied with the freshman mathematics courses. He does not state, however, the number who reported dissatisfaction with a general mathematics course.

Nearly a decade later, Adams[19] made a survey of the offerings in mathematics of twenty-six junior colleges in California, as indicated by their catalogues. He concluded that a typical junior college offering in mathematics consisted of "elementary algebra, elementary plane geometry, intermediate algebra, plane trigonometry, solid geometry, plane analytic geometry, differential calculus, integral calculus, solid analytic geometry and infinite series."

After calling attention to the similarity of the junior college offerings to the offerings of the lower division of the University of California, Adams suggests further uniformity by proposing

[17] J. W. Young, "The Organization of College Courses in Mathematics for Freshmen." *American Mathematical Monthly*, pp. 6-14, January, 1923. Also *American Mathematical Monthly*, p. 283, September, 1922.

[18] E. J. Hills, "Junior College Mathematics." *School Science and Mathematics*, 29:880-885, 1929.

[19] L. J. Adams, "Mathematics in California Junior Colleges." *Junior College Journal*, 7:194, 195, 1937.

the same name and number for the course, and a minimum list of essential topics in mathematics to be covered in both the junior college and the university. However, he does not show how this uniformity will lend itself more readily in adjusting a curriculum to the individual student's needs.

In 1939 Hannelly[20] reported, in his doctoral dissertation, a survey of the mathematical offerings as indicated in the catalogues of 352 junior colleges located in forty-four states. The topics, as they appeared in order of frequency, are algebra, trigonometry, analytic geometry, calculus, and "combination courses" for freshmen.

The report states that the "combination courses" usually consisted of topics selected from arithmetic, college algebra, trigonometry, and analytic geometry. After an examination of seventy-eight textbooks and four syllabi used in junior college mathematics, Hannelly presents his conclusions in a very brief but concise manner. Among his conclusions is that "most texts in combination-courses consist of consecutive treatment of the three regular freshman courses."

The study also included correspondence with fifty mathematics teachers in junior colleges, and it is pointed out that twenty of these teachers expressed dissatisfaction with the present mathematics courses. In evaluating his report, Hannelly proposes for those students not directly interested in mathematics a general mathematics course "unified on the bases of functionality, reality of the problems, and the social and scientific importance of these problems"[21] and that the construction of such a course include the cooperative effort of both the teacher and the student.

In the same year that Hannelly sought the opinion of mathematics teachers concerning their evaluation of the mathematics courses in the junior college, the Committee on the Improvement of Science in General Education, appointed by the Ameri-

[20] Robert J. Hannelly, "Mathematics in the Junior College." *The University of Colorado Studies, Abstracts of Theses for Higher Degrees*, pp. 65-68. Boulder, Colorado, 1939.

[21] *Ibid.*, p. 68.

can Association for the Advancement of Science, sent a questionnaire to colleges, universities, and teacher-training institutions. The questionnaire[22] requested information concerning the extent to which "the conventional introductory courses in mathematics, as represented by a majority of current textbooks" were meeting the requirements of the "non-specializing" student. Not only was there general dissatisfaction expressed in regard to the conventional introductory mathematics courses for the student not specializing in mathematics, but only 51 of the 136 instructors who answered the questionnaire indicated that they felt that the course could be "significantly" improved for this group of students.

A decade ago, a survey of published articles indicated a marked lack of interest in the improvement of college teaching[23] but now, if the interest in a subject may be judged by the published articles in that field, the interest in college mathematical education is suddenly on the increase. In fact, in one bibliography of mathematical periodical literature covering the period 1920–1941, more than half the references under "college mathematics" bear a publishing date falling within the last five years.[24] Thus mathematical publications seem to be reflecting increased interest in college mathematics education, and in particular interest in the mathematical education of the freshman. The cause for this added interest necessitates a brief survey of the rise of the general mathematics movement, and to give this is the purpose of the chapter which follows.

[22] J. S. Georges, "Humanizing the Curriculum of the Natural Sciences and Mathematics." *School Science and Mathematics*, Vol. 40, p. 454, May, 1940.

[23] Joseph Seidlin, *A Critical Study of the Teaching of Elementary College Mathematics*, p. 96. Bureau of Publications, Teachers College, Columbia University, New York, 1931.

[24] W. L. Schaaf, *A Bibliography of Mathematics Education*. Stevinus Press, Forest Hills, New York, 1941.

CHAPTER III

RETROSPECTIVE VIEW OF THE GENERAL MATHEMATICS MOVEMENT

THE TERM "GENERAL MATHEMATICS"

COLLEGE mathematics for freshmen is organized along two general lines—traditional compartmental and so-called modern non-compartmental. A common sequence of subjects in the compartmental organization is algebra, trigonometry, analytic geometry, and, in some cases, the calculus. However, the calculus is usually reserved until the sophomore year. The non-compartmental plan is represented by a two-semester course of mathematics consisting of topics from algebra, trigonometry, and analytic geometry, the calculus, and other branches of mathematics in harmony with the objectives of the course. These topics are usually correlated and built around some unifying theme, such as the historical development of the subject, function concept, etc. But as previous writers[1] have pointed out, general mathematics may be based on some main theme, such as statistics, functions, or mensuration; or it may be organized for some special group, such as agriculture or education students. On the other hand, it may consist only of those topics basic to an introductory course for students taking college mathematics. In any case, an attempt is not made to keep the subjects in separate air-tight compartments nor does this necessarily imply a forced correlation. An attempt is usually made to arrange the content material in a psychological rather than a logical order.

The textbooks in the non-compartmental organization may bear any of the following titles: General Mathematics, Correlated Mathematics, Fused Mathematics, Freshman Math-

---

[1] Frank Anderson, "The General Mathematics Course in Higher Institutions." Unpublished Master's thesis, University of Arizona, 1938.

ematics, Introductory Mathematics, Social Mathematics, etc.[2]

In harmony with other writers in the field[3] and the recommendation of the Joint Commission the term "general mathematics" in this study will denote the type of course just described. Similarly, emphasis in this report will be placed upon the general mathematics offered to the cultural terminal class of students, with little or no emphasis upon the general mathematics offered to special groups, such as those in agriculture, home economics, or business.

MATHEMATICS IN EARLY AMERICAN COLLEGES

Before further study is made of the general mathematics for the large academic group, it will be well to recall that in the early development of mathematics by the Arabs and Hindus no attempt was made to segregate arithmetic from algebra or geometry. Likewise, in the time of Euclid, mathematics was not separated into subjects.[4] Even in 1643 at Harvard, the mathematics curriculum was a combination of arithmetic and geometry offered in the last year. Ward's *Mathematics*, a textbook used at Harvard at the beginning of the seventeenth century, contains sections with special emphasis on arithmetic, algebra, geometry, and surveying, but there was no apparent attempt to

---

[2] See Bibliography in Appendix E.

[3] P. C. Scott, "A Comparative Study of Achievement in College Freshman Mathematics." Unpublished Doctor's dissertation, George Peabody College for Teachers, Nashville, Tennessee, 1930.

R. J. Hannelly, "Mathematics in the Junior College." Unpublished Doctor's dissertation, University of Colorado, Boulder, Colorado, 1939.

F. Anderson, "The General Mathematics Course in Higher Institutions." Thesis submitted to the faculty in the Graduate College, University of Arizona, 1938.

H. A. Wright, "An Evaluation of Certain Textbooks in General Mathematics for College Freshmen with a View to Formulating a Course Which Affords More Satisfactory Preparation for Calculus." Unpublished Doctor's dissertation, New York University, New York, 1932.

The Final Report of the Joint Commission of the Mathematical Association of America and the National Council of Teachers of Mathematics, *The Place of Mathematics in Secondary Education*. Bureau of Publications, Teachers College, Columbia University, New York, 1940.

[4] C. McCormick, *The Teaching of General Mathematics in the Secondary Schools of the United States*. Bureau of Publications, Teachers College, Columbia University, New York, 1929.

draw a sharp line of demarcation between any of the topics of mathematics. At the beginning of the eighteenth century, however, the separation of mathematics into its various branches may be observed by the fact that in the same institution arithmetic and algebra were offered as separate subjects in the freshman and sophomore year respectively. As these subjects were pushed down into the secondary schools the content of the courses became more standardized and the lines of separation more distinct.

The growth in mathematics during the first part of the nineteenth century was very slow. In higher mathematics there were few American mathematicians who were known abroad or who had made substantial contributions in the field. In the middle of the century, Benjamin Pierce at Harvard was the outstanding investigator in the field. However, at the close of the century, mathematics entered a period of rapid development. The International Commission in its report of 1911 describes the period thus: "The mathematical development in the United States during the last quarter of a century has been so unexpected and sudden, and so remarkable in its proportions, that it can scarcely be termed an evolution, but merits rather to be called a revolution. . . . Roughly speaking, the years 1880–1890 may be said to mark a new departure in the mathematical education of the United States."[5]

It was in the period 1880–1890 that James Joseph Sylvester came from England to Johns Hopkins University and encouraged work in higher mathematics, and that Harvard sent some of its best mathematics students abroad. It was during this period that Harvard reorganized the course in the calculus. The object of the new organization was to have a closer correlation between mathematics and physics. Thus, leaders in mathematics saw the beginning of a movement to break down the compartmental lines of their subjects.

[5] International Commission on the Teaching of Mathematics, The American Report Committee No. XII, *Graduate Work in Mathematics in Universities and in Other Institutions of Like Grade in the United States*, pp. 42-43. Government Printing Office, Washington, D. C., 1911.

COMMITTEE REPORTS

The idea of cutting across subject-matter lines in the organization of courses was expressed by the Committee of Ten in its report of 1892. The report indicated a step toward the general mathematics movement in that the committee was agreed that a change was needed in the teaching of mathematics and recommended "perspective views or broad surveys" of mathematics in the elementary schools.[6]

After emphasizing the need for the correlation of algebra, arithmetic, and geometry, President Eliot expresses the attitude of the Committee of Ten in these words: "Here is a striking instance of the interlacing of subjects which seem so desirable to every one of the nine conferences."[7] It is interesting to observe that in the Report suggestions of subject content do not extend beyond the twelfth grade.

The movement for the eradication of the boundary lines between subjects was discussed by the International Commission on the Teaching of Mathematics in detail as follows:

The recasting of the series of elementary subjects into a consecutive course in mathematics presents many interesting aspects. In a striking passage of his Elementarmathematik vom höheren Standpunkte aus, 1, 1908, Professor Felix Klein draws an interesting contrast between two systems of mathematical thought and development. In the first—
a particularistic conception of mathematical science is fundamental, dividing the whole into a series of well-defined provinces. In each of these one seeks to gain his ends with the minimum use of outside means or dependence on allied branches. The ideal is a structure of each division beautifully crystallized out and logically complete in itself.
The other system, on the contrary—
lays the chief emphasis on an organic interrelation of the separate branches, and on the numerous suggestions they afford each other. It prefers, accordingly, the methods which open up the simultaneous understanding of several branches from a unifying standpoint. Its

---

[6] Report of the Committee on Secondary School Studies Appointed at the Meeting of the National Education Association, July 9, 1892, p. 14. Government Printing Office, Washington, D. C., 1893.

[7] National Education Association, *Report of the Committee on Secondary School Studies*, p. 24. U. S. Bureau of Education, 1893.

ideal is the comprehension of all mathematical science as of one entity. . . .

The recent experiments in unifying the mathematical curriculum of the technological course represent a reaction against this prevalence of the first system distinguished by Klein. In discussing them it may be noted first that a similar tendency has some time since led to the introduction of the elements of analytical geometry in the form of graphs into the preparatory instruction in elementary algebra, and that there is a tendency in recent textbooks of elementary geometry to depart more widely from strict Euclidean traditions. The revision of the more advanced program has sometimes taken the partial form of merely eliminating college algebra or analytic geometry as a distinct subject. In one recent case the process has been carried further by combining the algebra, analytic geometry, a first course in calculus, and elementary differential equations into a single course and textbook.

Among the characteristics of this revision are:

The earlier introduction and consequently more prolonged study of the calculus ideas, and methods, function, derivative, etc.;

The treatment of analytic geometry rather as a general mathematical method than as a separate subject, with less emphasis on the conic sections and more on the geometric interpretation of a great variety of equations, including parametric forms;

The consecutive discussion of functions of one variable by the methods of algebra, analytic geometry, and calculus;

The subsequent combination of solid analytic geometry and partial differentiation in a briefer discussion of functions of two variables.

The working out of such a program is necessarily attended with some sacrifice of formal simplicity, and with occasional apparent loss of continuity. The advocates of it believe that such a thorough revision has, by testing values and eliminating non-essentials, effected considerable economies in time, and that the advantages of the plan will decidedly outweigh these defects, particularly as more prolonged experience facilitates the actual teaching. The student should for example gain power of discriminating attack by the use for each problem of the method best adapted to its solution. Such a program may prove less easy both to teacher and student, but may at the same time be none the less worthwhile. The chief difficulties to be guarded against are apparent or excessive discontinuity of subject matter, and the corresponding dispersion of interest and attention on the part of the student. He finds it difficult to grasp the unity of mathematical analysis and easy to lose the thread in passing from the methods of algebra to those of calculus and then of trigonometry. It must be borne in mind, however, that this sort of transition is just what must be easily made in any successful application of mathematics, and that the relative ease of completing a somewhat arbitrarily defined subject and then passing to the next may be too dearly purchased. The result of the experiments in this direction will certainly prove of much interest.[8]

EARLY FRONTIER LEADERS

E. H. Moore, in 1902, gave a notable address at the annual meeting of the American Mathematical Society in which he emphasized the need for the correlation of the different subjects of the curriculum.[9] The National Committee on Mathematical Requirements in 1923 suggested: "A convenient starting point for the history of the modern movement in this country may be found in E. H. Moore's presidential address before the American Mathematical Society in 1902."[10]

Frontier thinkers in Europe were also stressing the need for a change in the teaching of mathematics. This idea was expressed by an American mathematics committee in these words: "But the movement here is only one manifestation of a movement which is world-wide and in which very many individuals and organizations have played a prominent part."[11] Professor John Perry in particular was severely criticizing the teaching of mathematics in England.[12] His influence was strongly felt even in America through such leaders as E. H. Moore.[13]

Also, in Germany, Klein, one of the greatest mathematics specialists, had raised his voice in criticism by advocating the

[8] International Commission on the Teaching of Mathematics, The American Report Committee No. IX, *Mathematics in the Technological Schools of Collegiate Grade in the United States*, pp. 21, 22. U. S. Bureau of Education, Bulletin No. 9. Washington, D. C., 1911.

[9] E. H. Moore, "On the Foundations of Mathematics," *Bulletin of American Mathematical Society*, No. X, Vol. X, pp. 402-424, 1903.

[10] A Report by the National Committee on Mathematics Requirements under the Auspices of the Mathematical Association of America. *The Reorganization of Mathematics in Secondary Education*, p. vi, 1923.

[11] *Ibid.*

[12] John Perry and Others, *Discussion on the Teaching of Mathematics*. The Macmillan Company, New York, 1901.

[13] E. H. Moore, "On the Foundations of Mathematics. *Bulletin of the American Mathematical Society*, No. X, Vol. X, 402-424, 1903.

fusion of mathematical subjects with early and increased emphasis on the idea of function and derivatives.[14]

All these leaders help to create an interest in mathematics education. This is shown by the fact that three important educational associations were immediately formed, namely, the Association of Mathematics Teachers of New England, the Association of Mathematics Teachers of the Middle States and Maryland, and Central Association of Science and Mathematics Teachers.

General Mathematics in High School

The interest in the improvement of the teaching of mathematics is shown also by the fact that general mathematics was introduced in the Horace Mann High School for Girls,[15] and correlated mathematics at the Lincoln (Neb.) High School[16] and at the Chicago University High School,[17] at the beginning of the twentieth century.

A great impetus was given to the correlation of the different branches of general mathematics on the high school level by the experimentation of W. D. Reeve[18] at the University High School of the University of Minnesota. During the seven-year period 1915–1922, Dr. Reeve experimented with units of mathematics that would be suitable to the ninth and tenth grade.[19]

During the next decade the idea of the correlation of mathe-

[14] International Commission on the Teaching of Mathematics, The American Report Committee No. IX, *Mathematics in Technical Schools of Collegiate Grade in the United States.* U. S. Bureau of Education, Bulletin No. 9. Washington, D. C., 1911.

[15] "The Mathematics Course in the Horace Mann High School." *Teachers College Record,* 7:88-103, 1906.

[16] Edith Long, "Correlation of Algebra, Geometry, and Physics." *Educational Review,* 24:309-311, 1902.

[17] G. W. Myers, *First-Year Mathematics for Secondary Schools.* University of Chicago Press, 1909.

[18] W. D. Reeve, "The Case for General Mathematics." *The Mathematics Teacher,* 15:381-391, 1922.

W. D. Reeve, *A Diagnostic Study of the Teaching Problems in High-School Mathematics.* Ginn and Company, Boston, 1926.

[19] R. Schorling and W. D. Reeve, *General Mathematics,* Book I. Ginn and Company, Boston, 1919.

W. D. Reeve, *General Mathematics,* Book II. Ginn and Company, 1922.

matics with other subjects declined and the movement for correlation of the compartmental subjects of mathematics
grew.[20] The general mathematics movement that was gaining headway in high school soon began to be advocated by
teachers of mathematics on the college level—especially for
freshman students.

Factors Influencing the General Mathematics Movement

Several factors aided in the furtherance of the general mathematics movement in college, in addition to the Report of the
Committee of Ten, the influence of John Perry, Felix
Klein, and E. H. Moore, and the general mathematics movement in high school. First, a new philosophy of education was
replacing the old; second, there was growing dissatisfaction with
the mathematics curriculum; third, there was a general trend in
curriculum revision and re-evaluation; fourth, the enrollment
in colleges was increasing; fifth, the rise of the junior college
movement was causing a re-examination of the freshman curriculum; and sixth, the writings of the advanced leaders were
beginning to modify the current thought.

A New Psychology and Philosophy

A new psychology of learning and philosophy of education
were being formed in the minds of educators. Study for the
sake of mental discipline was passing out of the school picture.
Converting the mind into a storehouse of facts for future use
was being frowned upon. Discredit had come upon those older
ideas of education which had been described by President John
Coulter, in 1891, when he said: "Evidently, the greatest, widest
truth is that the mind is to be made powerful by exercise and
it will always be a secondary consideration whether this exercise shall come by loading the memory with the words and forms
found in several languages . . . or shall come by a direct study

[20] C. McCormick, *The Teaching of General Mathematics in the Secondary
Schools of the United States*, p. 26. Bureau of Publications, Teachers College,
Columbia University, New York, 1929.

of facts, and causes and laws, as found in science and history and literature." [21]

The International Commission of the Teaching of Mathematics reported that there was a tendency at this time (1911), first, to provide, by using applied problems, correlation between mathematics and engineering subjects; and second, to organize the subject matter of the mathematics curriculum "according to its difficulty or its needfulness as an auxiliary, with little or no reference to the traditional division of subject matter, into algebra, analytic geometry, and calculus." [22]

An attempt was being made to disregard compartmental subject lines and to organize the subject matter to meet the needs of the individual without reference to the subject itself. This viewpoint gained great momentum after the close of the first World War. Hopkins describes it as follows:

When some one in the year 2000 writes the history of American education for the twentieth century, the decade between the close of the World War and the financial and economic collapse which heralded the great depression will stand out as of peculiar importance. It was in these years that the great battle of educational ideas took place. The death struggle between two opposing types of curriculum practice was fought and decided. On the one side was the large group of educators who championed the subject curriculum; on the other was the small group of educators who advocated the experience curriculum. A decision was rendered in 1929. The social and economic events immediately following the depression caused educators to stop, look, listen, think, and evaluate the practices of the preceding decade. As a result, from the kindergarten through the liberal-arts college the subject curriculum with its basic educational ideas began to decline, and the experience curriculum with its fundamental principles began to increase. The rapid acceleration which began in 1931 has in this year, 1937, almost reached a tidal wave.[23]

[21] J. M. Coulter, President Elect, *Practical Education*, p. 9. (An address before the students of Indiana University). Carlon and Hollandback, Printers, 1891.

[22] International Commission of the Teaching of Mathematics, Annual Report; Committee No. IX, *Mathematics in the Technical Schools of Collegiate Grade in the United States,* p. 18. U. S. Bureau of Education, Bulletin No. 9. Washington, D. C., 1911.

[23] L. T. Hopkins, *Integration—Its Meaning and Application*, p. 197. D. Appleton-Century Company, Inc., New York, 1937.

Dissatisfaction with the Mathematics Curriculum

Concurrently with the change in philosophy there was a growing dissatisfaction with the organization and teaching of mathematics. A low degree of mastery was being exhibited by the average student, and the per cent of failures was increasing in the elementary school,[24] in the high school,[25] and in the college.[26] The value of the traditional mathematics was questioned for the student who does not specialize in a science that requires it as a tool. The idea that the students lacked a mastery of the simple mathematical techniques was not an idle dream on the part of the critics of mathematics. It is reported that on an examination given to 1,700 freshmen, 50 per cent of these students could not find the value of three-fourths divided by two-thirds. The objection may be raised that such operations have not been used in their high school work (at least it appears that *this* operation could not have been used *correctly*) and that the emphasis of high school mathematics is the understanding of the solution of equations, whether the mathematics is the traditional algebra or general mathematics. However, investigators have found that over 60 per cent of college freshmen cannot substitute 2 for a and 3 for b in the equation $x = \frac{1}{2}ab^2$, and obtain the correct result. Half of the same group could not solve $y = m^2 + n^2$ if $m = 3$ and $n = 4$.[27] We cannot claim that they have been spending time mastering so-called non-essentials, such as square root and radicals, when tests show that more than half could not approximate $\sqrt{7}$ to the nearest tenth nor express $\sqrt{X^6}$ as a power of X. The impression should not be left that students have not until recently experienced difficulty with courses in mathematics. In 1903,

[24] R. L. Morton, *Teaching Arithmetic in the Intermediate Grades*, p. 7. Silver Burdett Company, New York, 1939.

[25] Jack Wolfe, "Mathematics Skills of College Freshmen in Topics Prerequisite to Trigonometry." *New York City, Thirty-ninth Annual Report of Superintendent of Schools*, pp. 259, 260, 1936-37.

[26] R. Strang, *Personal Development and Guidance in College and Secondary Schools*, p. 167. Harper and Brothers, New York, 1934.

[27] L. C. Pressey, "The Needs of Freshmen in the Field of Mathematics." *School Science and Mathematics*, pp. 238-243, March, 1930.

Professor James M'Clure[28] in an address before the University Conference at Vanderbilt University pointed out that 44 out of his classes of 142 failed to pass the course. The cause of the large number of failures was given as "insufficient preparation to meet the requirements of the freshman class." However, it should be noted that two of the deficiencies of these entering freshman students were "inability to complete correctly the square of a quadratic" and "lack of knowledge of the log series," both of which may indicate that the standard of achievement was higher for the freshmen of 1903 than it is now. In any case the criticism of the teaching of mathematics because of the large number of failures has become pronounced in recent years.

Barnett says, "The universities, our own Ohio College, for example, have found their incoming students woefully deficient in their ability to cope with elementary arithmetic and algebraic technique, as well as with simple common-sense problems." [29] A study made under the auspices of the Texas Section of the Mathematics Association of America reveals that "Reports from Colleges and Universities show failures ranging in general from 25 to 50 per cent of class enrollments." The committee reports that "During the past several years the percentage of failures in college freshman mathematics has become alarming. . . . It is significant to note that the percentages of failures are rapidly increasing from year to year. So noticeable has this become that studies are being undertaken by individuals and committees with the hope of determining the causes, and if possible, find remedies to alleviate the unfortunate situation." [30]

It should be observed that the cause of the failure does not seem to be inherent in mathematics itself. Those who fail in satisfactorily completing a course in mathematics are not likely

---

[28] James M'Clure, "Failures in Freshman Mathematics." Paper read before the University School Conference, Vanderbilt University, May 1, 1903. *Vanderbilt University Quarterly*, Vol. III, No. 4, October, 1903.

[29] J. A. Barnett, "A Proposal for the Improvement of Teaching Mathematics." *National Mathematics Magazine*, p. 74, December, 1934.

[30] J. M. Bledsoe, "Failures in College Freshman Mathematics." *Texas Outlook*, p. 18, October, 1940.

to be among those who are in the upper range of ability in their other subjects. It has been shown[31] that of those students who received an *F* in one subject, less than one in thirty received an *A* in any course. Of the 1,676 students included in the study about 18 per cent failed in mathematics and five out of six of these also failed in some other subject. Guggenbuhl contends that the person who is good in everything except mathematics is fiction. The deficiencies do not seem to be confined to mathematics nor do the types of errors seem to be many in number. In fact the common errors made by students in their mathematics work through the calculus have been listed. This list contains only 120 errors, which are mostly those in the elementary phase of mathematics.[32] Others[33] have pointed out that the teaching problems caused by extreme weakness in mathematical ability have become more acute within the past few years. Perhaps this has been one of the contributing factors in causing a decrease in the mathematics required for graduation at some colleges. It has been reported that in 35 per cent of certain "Southern" colleges, both large and small, mathematics is not required for the bachelor's degree, and in the case of the teachers colleges 68 per cent do not require mathematics for graduation. In Northern colleges the required mathematics was found to be even less than that in the South.[34]

[31] Laura Guggenbuhl, "The Failure in Required Mathematics at Hunter College." *The Mathematics Teacher,* 30:68-75, February, 1937.

[32] Alan D. Campbell, "Some Mathematical Shortcomings of College Freshmen." *The Mathematics Teacher,* 27:420-425, December, 1934.

[33] Ina Holroyd, "Weaknesses of High School Students Who Enter College Mathematics and a Suggested Remedy." *The Mathematics Teacher,* 27:128-137, March, 1934.

Laura Guggenbuhl, "The Failure in Required Mathematics at Hunter College." *The Mathematics Teacher,* 30:68-75, February, 1937.

E. A. Cameron, "A Program in Freshman Mathematics Designed to Care for a Wide Variation in Student Ability." *American Mathematical Monthly,* 47:471-473, August, 1940.

A. D. Campbell, "Some Mathematics Shortcomings of College Freshmen." *The Mathematics Teacher,* 27:424-425, December, 1934.

C. C. Richtmyer, "The Functional Mathematics Needs of Teachers." *Journal of Experimental Education,* June, 1938.

[34] P. M. Ginnings, "Mathematics and Science Requirements for the Liberal Arts Degree in Southern Colleges." *High School Quarterly,* 23:10-12, October, 1934.

The Trend in Curriculum Revision

In addition to expressing dissatisfaction with the results of mathematical education the educator and mathematician have attempted to re-evaluate and reorganize the curriculum. It is not necessary to discuss the countless courses of study that have been made; one glance at the stacks of courses of study in the average university[35]—one university alone contains 84,653 courses of study—will remind one of the tremendous effort expended. The result of this tendency on subjects offered in the first two years of college has been adequately pointed out.[36]

Increased Enrollments

The increased school enrollment had a decided effect upon the organization of the curriculum and the methods of teaching. No longer are the colleges preparing students for only the three professions of medicine, law, and theology. They are preparing students to take their place as citizens in all walks of life. The increase in the population 18–21 years of age was only 63 per cent during the period 1900–1938, but the increase in college enrollments during the same period was 468 per cent. As Seashore has said: "A college education is more common now than a high school education was a few years ago. . . . The doctorate is more commonplace now than a bachelor's degree was fifty years ago. There are now more postdoctorate students in the universities than there were candidates for the doctorate at the beginning of the century. . . . So the surging waves of increasing demands for higher education roll against the walls of the halls of learning. These walls are bulging and trembling under the pressure of mass education." [37] The growing dissatisfaction of this multitude has led to a reorganization of the curriculum content and methods of teaching.

[35] H. B. Bruner and Others, *What Our Schools Are Teaching*, p. 9. Bureau of Publications, Teachers College, Columbia University, New York, 1941.

[36] A Report on Problems and Progress of the General College. Prepared by Staff of the General College, Malcolm S. MacLean, Director, "Curriculum Making in the General College." University of Minnesota, June, 1940.

[37] C. E. Seashore, *The Junior College Movement*, pp. 3, 4. Henry Holt and Company, New York, 1940. Quoted by permission of the publisher.

Not only have the organization and content been changed in an endeavor to prepare the student to realize the goals he has set for himself, but attempts have been made to improve methods of teaching. Classroom teaching has been critically evaluated by the mathematician[38] and experimentation is taking place. In August, 1941, there appeared in the *American Mathematical Monthly* an article entitled "A Program in Freshman Mathematics Designed to Care for a Wide Variation in Student Ability" by E. A. Cameron,[39] which describes the ability grouping in freshman mathematics that is being attempted at the University of North Carolina. In an article[40] in the same magazine shortly afterwards H. L. Dorwart gives the following comment: "With several minor changes, the article 'A Program in Freshman Mathematics Designed to Care for a Wide Variation in Student Ability' by E. A. Cameron, . . . might have referred to Washington and Jefferson College instead of to the University of North Carolina." He then describes the attempt at ability grouping at Washington and Jefferson College. It is interesting to notice that the present experiment in ability grouping was begun in both colleges in 1938, that the same tests for classification are used in each, and that both report a reduction in failures. Members of the mathematics faculty of Queens College (New York City) reported to the author a similar experience. In fact, the result of a recent questionnaire sent to forty-eight colleges showed that, of the forty-three institutions which replied, twenty-seven had tried ability grouping. Twenty-two indicated that they were in favor, four disapproved, and seven did not have sufficient evidence since the innovation to form an opinion. Nearly all the grouping was in the elementary mathematics for the freshman and sophomore years. The ques-

---

[38] J. Seidlin, *A Critical Study of the Teaching of Elementary College Mathematics.* Bureau of Publications, Teachers College, Columbia University, New York, 1931.

[39] E. A. Cameron, "A Program in Freshman Mathematics Designed to Care for a Wide Variation in Student Ability." *American Mathematical Monthly,* 47:471-473, August, 1940.

[40] H. L. Dorwart, "Comments on the North Carolina Program in Freshman Mathematics." *American Mathematical Monthly,* pp. 37-39, January, 1941.

tionnaire indicated that "A definite trend toward ability grouping is evident in the State Universities."[41]

RISE OF THE JUNIOR COLLEGE

Another factor that had a tremendous bearing upon the development of the general mathematics movement was the rise of the junior college. The acceptance of the junior college as a part of the secondary school necessitated the redefining of the term "secondary school." The concept was now enlarged to include grades seven to twelve, as was the view of the National Committee on Mathematics Requirements,[42] to grades seven to fourteen, as expressed by the final report of the Joint Commission of the Mathematical Association of America and the National Council of Teachers of Mathematics.[43]

That junior college mathematics was being accepted as part of the secondary school is indicated by the following quotation from a committee report submitted to the International Commission in 1911: "The fact that the men who are preparing in the schools of education for this college work usually follow at the same time the courses offered on the teaching of secondary mathematics shows that they realize that the problems of teaching in the first years of college are almost identical with those in the last two years of the high school. France and Germany long ago fully appreciated this when they included in the courses of study for their secondary schools the mathematics which we usually teach in the first two college years." [44]

However, the junior college was not accepted in general at

---

[41] R. O'Quinn, "Status and Trends of Ability Grouping in the State Universities." *The Mathematics Teacher,* 38:215, May, 1940.

[42] A Report by the National Committee of Mathematics Requirements under the auspices of the Mathematics Association of America, *The Reorganization of Mathematics in Secondary Education,* 1923.

[43] Report of the Joint Commission of the Mathematical Association and The National Council of Teachers of Mathematics, *The Place of Mathematics in Secondary Education.* Bureau of Publications, Teachers College, Columbia University, New York, 1940.

[44] Report submitted to the International Commission on the Teaching of Mathematics by subcommittee, C. B. Upton, Chairman, "The Training of Teachers of Mathematics in Professional Schools of Collegiate Grade." Reprint from *Educational Review,* p. 391, April, 1911.

this time as a part of the secondary school. In 1929 McCormick did not include junior college mathematics as a part of the secondary school program. Much has been written setting forth the arguments for and against the acceptance of the junior college as a part of the secondary school; and although the debate has not ended, there seems to be a tendency to accept the junior college as a part of the general education program which includes the secondary schools.[45]

The rise of the junior college has been due in general to the rise of the educational level of the masses, the vocational demands of technology, increase in leisure time and unemployment of American youth, and educational economy during periods of financial depressions. The per cent of youth in the United States who are attending secondary schools has more than doubled since 1880. "We are told that more pupils are today enrolled in American secondary schools than are enrolled in all other nations combined." [46] In like manner the momentum of the movement for the education of the population has carried over into the colleges. Illiteracy is disappearing from our country. The desire of the parent is to have the child educated so that he won't have to work as his father had to work. The great mass of young people come to college not to prepare for the vocation of their parents but for a vocation or profession which they feel will permit them to have a higher standard of living. Every teacher is aware of this urge of the youth to seek a profession that will provide an income greater than the income of his parents. This tendency to seek a higher vocational level is encouraged by the parents. The junior college provides, in part, an answer to that demand. Gradually, as the masses realized that it was impossible for every child to be educated for a profession and individual parents saw that it was not possible for "my child" to enter the coveted "white collar" field, the urge for the higher position was released in the cry for vocational training.

[45] W. C. Eells, *The Junior College,* Chapter XXIV. Houghton Mifflin Company, Boston, 1931.

[46] C. E. Seashore, *The Junior College Movement,* p. 3. Henry Holt and Company, New York, 1940. Quoted by permission of the publisher.

However, the junior college was not established for the purpose of vocational training, according to Wilson.[47] He points out that the vocational aims were incidental to the "cultural" purposes. Others indicated their agreement with this view by stating: "The University Junior College is designed primarily to provide broadened intellectual training to that large body of students who seek an overview of modern life and of man's activities, rather than specialized training." [48] However, the sentiment for specialized training increased and now the enrollment in the vocational curriculum excels all others in the junior college. This is expressed by Seashore in these words: "The advance of technology demands the privilege of technological education for the workers." [49] The attempt to meet these educational objectives led to the establishment of terminal survey courses in cultural background subjects and specialized electives fundamental to certain trades and vocations.

The increase in the leisure time of the common man was a contributing factor to this demand for increased educational opportunities. With the rise of technology came the shorter working day with more hours for recreation, and with the increase of unemployment came hours of unspent energy and millions of idle hands. These extra free hours which made it possible for adults to continue their education gave impetus to the junior college adult education movement.

We should not overlook the fact that finance also played an important part in the establishment and success of the junior college. As the college enrollments increased, larger physical plants and teaching and administrative staffs were required so that with many colleges there was reached a point of diminishing returns. The opportunity to be relieved of a portion of the large influx of freshmen and sophomores was welcomed. To the parent it meant less expense in the days of the depression.

---

[47] T. H. Wilson, "The First Four Year Junior College." *Junior College Journal,* 9:365, April, 1939.
[48] Report by the Staff of the General College, *Curriculum Making in the General College,* p. 13. University of Minnesota, Minneapolis, Minnesota, 1940.
[49] C. E. Seashore, *Junior College Movement,* p. 7. Henry Holt and Company, New York, 1940. Quoted by permission of the publisher.

It was possible for the student to remain at home and commute to school and, in many cases, continue with some part-time job in the home or local community.

The result of a recent survey shows that more than half of the students answering the questionnaire attended their respective junior college because of its location.[50] Thus, educational economy, unemployment with the increase of leisure time, rise of technology and the growth of education among the people all contributed to the birth and rapid development of the junior college.

President William R. Harper, first president of the University of Chicago, has been called the "Father of the Junior College Movement." [51] However, there were attempts and suggestions in regard to the segregation of freshmen and sophomores as early as 1852.[52] In 1869, W. W. Folwell, then president of the University of Minnesota, advocated the transfer of the first two years of college work to the secondary schools.[53] The administrators of the Universities of Illinois, Pennsylvania, and Michigan Western Reserve discussed the possibility of differentiating between the work of the upper and lower classmen, and some attempt was made but gained little headway. Eells points out, however, that the work at the University of Michigan was the "direct seed from which later sprang the extensive junior college development in California."[54]

In 1892, the University of Chicago was reorganized under President Harper. The freshman and sophomore years were organized into the "academic college" which was later (1896) renamed the "junior college." Eells states that "As far as known this is the first use of the term 'junior college.' " [55] During this period, when the colleges were making a distinct division be-

---

[50] *Ibid.*

[51] F. M. McDowell, *The Junior College*. U. S. Bureau of Education, Bulletin No. 35. Washington, D. C., 1919.

[52] Walter C. Eells, *The Junior College*, p. 45. Houghton Mifflin Company, Boston, 1931.

[53] W. W. Folwell, *University Addresses*, pp. 37-38. The H. W. Wilson Company, Minneapolis, 1909.

[54] Eells, *op. cit.*, p. 46.

[55] *Ibid.*, p. 47.

tween the upper and lower years of study, high schools were beginning to lengthen their period of instruction. Perhaps the first school to extend its work beyond the customary secondary curriculum, but including only the freshman courses, was a Catholic school established in 1677 at Newton, Maryland. Eells says, "It might be called the earliest junior college, since in addition to secondary work it did not carry its students beyond the freshman year in college." [56] There were certain seminaries, particularly for women, which offered a type of junior college work. Lasell Seminary was offering such a type of work in 1859.

However, the real birth of the junior college movement came with the establishment of the first public junior colleges. In 1895, East Side High School of Saginaw, Michigan was giving a year's work equivalent to that given in the freshman year at the University of Michigan.[57] About this same time a high school in Joliet began to offer an extra year of courses that permitted their students to enter the University of Illinois as sophomores.

Under the influence of President Harper of the University of Chicago, Joliet organized the first public junior college, which is still operating.[58] At this same time, The Junior College at Bradford, Massachusetts, was established.[59] By 1904 there were reported eighteen junior colleges, but the growth was slow. Many of these junior colleges were in existence only a short period. According to the United States Bureau of Education, by 1918 there were forty-six junior colleges and in 1928 there were 248.[60] No doubt this survey was not complete but it does indicate the growth during this ten-year period.[61] The rise of the junior college was mainly from two sources. Some four-year

[56] *Ibid.,* p. 57.
[57] *Ibid.,* p. 53.
[59] *Ibid.,* p. 54.
[59] U. S. Office of Education, *Junior Colleges,* p. 43. Bulletin 1936, No. 3. Washington, D. C., 1936.
[60] U. S. Office of Education, *Statistics of Universities, Colleges, and Professional Schools,* p. 1. Bulletin 1929, No. 38. Washington, D. C., 1930.
[61] W. C. Eells, *The Junior College,* Chapter 29. Houghton Mifflin Company, Boston, 1931.

institutions had taken the advice of President Harper and had contracted their programs.[62] On the other hand, many new private and public junior colleges had been established. For a time the public institutions were rapidly gaining over the private.[63] The reports of McDowell in 1917,[64] Koos in 1923,[65] Whitney in 1928,[66] Campbell in 1929,[67] and the American Association of Junior Colleges in 1930,[68] all indicate the accelerating growth of the two-year institution during this period.

The next decade, 1930–1940, saw the growth of an educational institution the like of which had never been witnessed here or abroad. Seashore describes it as follows: "The junior college movement is perhaps the most significant mass movement in higher education that this or any other country has ever witnessed in an equal period of time." [69] Growth was so great that in 1937 provision was made for a "junior college section" by the Mathematical Association of America and The National Council of Teachers of Mathematics in their annual meetings.

The *Junior College Directory*[70] for 1941 lists 610 institutions as public or private junior colleges with an enrollment of 236,162. Although the number of private institutions is still in excess of those publicly controlled, the public junior college enrolls 71 per cent of all the junior college students.[71]

[62] U. S. Office of Education, *Junior Colleges*, p. 19. Bulletin 1936, No. 3. Washington, D. C., 1936.

[63] U. S. Office of Education, *Statistics of Universities, Colleges, and Professional Schools*, p. 1. Bulletin 1929, No. 38. Washington, D. C., 1930.

[64] F. M. McDowell, *The Junior College*, p. 35. U. S. Bureau of Education, Bulletin 1919, No. 35. Washington, D. C., 1919.

[65] L. V. Koos, *The Junior College Movement*. Ginn and Company, Boston, 1925.

[66] F. L. Whitney, *Junior College in America*. Colorado State College of Education, Greeley, Colorado, 1928.

[67] D. S. Campbell, "Directory of the Junior College, 1932." *The Junior College Journal*, Vol. II, No. 4, pp. 235-248, January, 1932.

[68] W. C. Eells, *The Junior College*, Chapter XXIX. Houghton Mifflin Company, Boston, 1931.

[69] C. E. Seashore, *The Junior College Movement*, iii. Henry Holt and Company, New York, 1940. Quoted by permission of the publisher.

[70] W. C. Eells and P. Winslow, *Junior College Directory, 1941*. American Association of Junior Colleges, Washington, D. C., 1941.

[71] W. C. Eells, *Present Status of the Junior College Terminal Education*, p. 9. American Association of Junior Colleges, Washington, D. C., 1941.

Eells in his report for the Commission on Junior College Terminal Education in 1941 said, "While it required the first twenty years of the century to develop the first hundred of the present group of junior colleges, each succeeding five-year period has added another hundred or more to the total number and there seems to be no tendency as yet for this rate of growth to diminish." [72] At the same time the average size of the junior college seems to be increasing, a growth which is fortunate since nearly a third of the present junior colleges annually enroll less than one hundred students.

These facts present phases of the growth of the heterogeneous movement in the United States which has many variations in curriculum and organization. In general the objectives of the junior colleges have been twofold. First, they have acted as preparatory schools for the upper divisions in the colleges and universities. Second, they have endeavored to give a terminal education. In fact, out of thirty-five separate statements of functions of the junior college the terminal and preparatory purposes outrank by far any other purpose as stated in 347 published articles and 343 junior college catalogues. [73]

The tendency of the early junior college was to stress the preparatory function of the institution. Of course, there were exceptions, such as Lasell Female Seminary, whose course of study in 1875 was largely terminal. [74] Certain junior colleges in California had made attempts to incorporate vocational curricula. [75] McDowell's report in 1918 indicated that 18 per cent of the courses in public junior colleges and only 9 per cent in the private junior colleges [76] could be classified as vocational. At the first junior college conference held in St. Louis in 1920 one of the topics discussed was "Junior College as Completion

[72] *Ibid.*

[73] Doak S. Campbell, *A Critical Study of the Purposes of the Junior College,* pp. 18-30. George Peabody College tor Teachers, Nashville, Tennessee, 1930.

[74] T. H. Wilson, "The First Four Year Junior College." *Junior College Journal,* p. 365, April, 1939.

[75] W. C. Eells, *Present Status of Junior College Terminal Education,* p. 21. American Association of Junior Colleges, Washington, D. C., 1941.

[76] F. M. McDowell, *The Junior College.* U. S. Bureau of Education, Bulletin 1919, No. 35. Washington, D. C., 1919.

School." [77] In regard to the discussion of this topic at that conference Eells states: "It would appear that recognition of the terminal function, therefore, at this period existed more as aspiration in the minds of the administrators than as realization in the experience of students and their parents." [78]

The reports of Koos,[79] Hollingsworth, and Eells in 1930,[80] and Colvert in 1937,[81] indicate a definite trend in the increase of the terminal offerings. Although there may not be a definite correlation between the increase in offerings and the increase in number of students taking the courses offered, the trend indicates that the interest in this type of work is increasing. A report in 1931 covering more than sixty junior colleges states that 20 per cent of the 15,000 students concerned were enrolled in a terminal curriculum.[82] A report in 1939 states that of 121,573 students in 426 junior colleges, 34 per cent were enrolled in terminal courses.[83] Similar studies in the registration of junior college students indicate that terminal courses are increasing in number and enrollment.[84]

Eells states that terminal education is just beginning to be recognized as important, and the increasing interest may be measured by the increase of literature in the field. In his recent bibliography, "The Literature of Junior College Terminal Education," almost half the titles were published after 1935.

In general the terminal courses have tended to group themselves into two categories—the terminal cultural courses and

[77] G. F. Zook, *National Conference on Junior Colleges, 1920.* U. S. Bureau of Education, Bulletin No. 19. Washington, D. C., 1922.

[78] W. C. Eells, *Present Status of Junior College Terminal Education,* p. 18. American Association of Junior Colleges, Washington, D. C., 1941.

[79] L. V. Koos, *The Junior College Movement,* p. 33. Ginn and Company, Boston, 1925.

[80] W. C. Eells, *The Junior College,* p. 489. Houghton Mifflin Company, Boston, 1931.

[81] C. C. Colvert, *The Public Junior College Curriculum,* p. 140. Louisiana State University Press, University, Louisiana, 1939.

[82] L. V. Koos and F. J. Weersing, *Secondary Education in California: A Preliminary Survey,* p. 91. State Department of Education, Sacramento, California, 1929.

[83] W. C. Eells, *Present Status of Junior College Terminal Education,* Chapters V, VI. American Association of Junior Colleges, Washington, D. C., 1941.

[84] *Ibid.,* pp. 24-25.

the terminal vocational courses. In a study of the enrollment of over forty thousand students, more than one third of the students were enrolled in a terminal business course.[85] It is not surprising to find courses in mathematics being developed for this class of students. Some of the texts for these students are *The Mathematics of Business*,[86] *Business Arithmetic for College Students*,[87] and *Business Mathematics*.[88] These courses usually consist of topics selected in harmony with the vocational interest of a special group of students.

In the same study, four per cent of the group were enrolled in terminal courses in agriculture. Here again we find courses being developed by the selection of certain topics of mathematics that seem to lend themselves readily to the vocational interests of this special group.[89] Nearly a third of the 40,000 students concerned in the investigation were enrolled in general cultural or public service terminal curricula.

In providing for those students who are preparing for further collegiate work, the junior college has fashioned a curriculum similar to that of the first two years of the university. In mathematics the usual offering in the freshman year is either algebra, trigonometry, and analytic geometry, or a two-semester course in general mathematics. The textbooks, in general, are those used by universities. Two thirds of the students are reported to be enrolled in this type of preparatory course, but less than one fourth of the freshmen enter the upper college division. Thus, it would seem that two fifths of the students are pursuing courses that are not designed to meet their needs.[90] Perhaps it

[85] *Ibid.*, pp. 51-52.

[86] H. E. Stelson, *The Mathematics of Business*. Houghton Mifflin Company, Boston, 1940.

[87] William S. Schlauch, *Business Arithmetic for College Students*. F. S. Crofts and Company, New York, 1939.

[88] Isaiah L. Miller, *Business Mathematics*. D. Van Nostrand Company, Inc., New York, 1935.

[89] H. B. Roe, D. E. Smith, W. D. Reeve, *Mathematics for Agriculture and Elementary Science*, Ginn and Company, Boston, 1928.

[90] W. C. Eells, *Present Status of Junior College Terminal Education*, p. 25. American Association of Junior Colleges, Washington, D. C., 1941.

W. C. Eells, *Why Junior College Terminal Education?* American Association of Junior Colleges, Washington, D. C., 1941.

can be safely said that the junior college enrollment will decrease in the preparatory curriculum and increase in the terminal curriculum. With the increase in terminal education enrollment may be an increase in the selection of terminal mathematics courses rather than college preparatory courses.

For the large academic terminal group of students we find developing a "fused" or general mathematics course. This course consists of selected topics of mathematics that are deemed of general or cultural interest. The topics are usually built around some unifying thread of interest and are presented from the psychological point of view rather than in the logical order.

VOICE OF THE LEADERS

The rise of the junior college movement was contemporary with an increased agitation on the part of leaders in the field for a change in the organization and teaching of mathematics. The closer correlation of mathematics and physics in the reorganization of certain mathematics classes by the Harvard mathematics faculty in the middle of the nineteenth century was an indication of dissatisfaction among certain leaders, but the feeling did not gain headway until the twentieth century. The address of E. H. Moore seemed to crystallize the idea that had been vaguely trying to emerge in the minds of the frontier thinkers. Slowly others[91] in the field began to advocate a new emphasis in mathematics. During the decade following Mr. Moore's address, college staffs began to reorganize mathematics instruction. The American Commissioners of the International Commission on the Teaching of Mathematics state:

Some institutions have found it convenient to rearrange in a measure the topics treated in the elementary courses (college mathematics), and the calculus is now sometimes begun in the freshman year, the more difficult parts of analytic geometry being postponed

[91] W. F. Osgood, "The Calculus in Our Colleges and Technical Schools." *Bulletin of the American Mathematical Society*, Vol. 13, p. 449, 1906-07. Also Symposium on "Mathematics for Engineering Students," a succession of papers in *Science*, Vol. XXVIII, July 17–September 4, 1908.

to the following year. The plan is often referred to as one of "fu-sion"—amalgamation—into a single homogenous course subjects hitherto treated in distinct courses; and we have heard much of "water-tight compartments." What has actually taken place in such institutions, however, has been a modification in the order in which the various topics of the mathematics of the first two years of the college and the technological schools are treated. The individual topics are studied in the main as they were formerly, an individual chapter in analytic geometry, for example, not being disintegrated and the unity of its method not being destroyed; but when this chapter is finished, the next topic may be a chapter in algebra or the calculus. The early application of the calculus to curve plot-ting, together with a broadening of the range of curves studied, and the introduction at an early stage of the approximate solution of numerical equations, both algebraic and transcendental, by graphi-cal methods reinforced by the calculus, are two important results of such a rearrangement of topics.[92]

In order to improve the teaching of college mathematics, method courses were advocated for those students who expected to teach on the college level. In 1911 one committee observes that "students are beginning to realize that there are important pedagogical problems to be solved, especially in connection with the teaching of the first year of college mathematics where, in general, more poor instruction is found than in any other place in our educational system, a situation which is easily ex-plained, for as a rule the teachers in the first year of the col-lege course are young men, fresh from the study of higher math-ematics, who have never had experience in teaching and who have never so much as discussed the most elementary topic in re-lation to such work."[93] An indication of the slowness with which the college revision in mathematics had taken place is the description of the "Changes Still to Be Desired" in the re-port of Committee No. X to the International Commission on the Teaching of Mathematics entitled "Undergraduate Work

[92] U. S. Bureau of Education, *Report of the American Commissioners of the International Commission on the Teaching of Mathematics,* pp. 41, 42. Bulletin 1912, No. 14. Washington, D. C., 1912.
[93] C. B. Upton, "The Training of Teachers of Mathematics in Professional Schools of Collegiate Grade." *Educational Review,* Vol. XLI, p. 391, April, 1911.

in Mathematics in Colleges of Liberal Arts and Universities." [94]
Their report states:

No well-defined notion of the changes still to be desired in the
directions thus far discussed exists in the minds of the average
mathematical staff. The majority of those formulating any ideal
express it as involving still further enlargement of the teaching
staff to enable more courses to be given, . . . and to make possible
the giving of all mathematical instruction to divisions of small
size.[95]

The suggestion was also made that "more frequent depart-
mental meetings" be held.

One noteworthy thing is this, that branches of modern mathematics marking successive steps in the progress of this science still
remain isolated as subjects of instructions, separate also from the
older theories. Little blending has yet occurred or confluence of
these tributaries into a homogeneous stream. Only within a decade
is there found any indication of purposeful progress toward the
combination and consolidation of accidentally severed lines of
mathematical thinking. Probably not until Descartes, Newton, Lo-
bacevsky, Galois, and Hamilton become dim names in the mists of
antiquity will the characteristic thoughts of all pervade the whole of
mathematical instruction. Yet toward such an end every college
teacher of the present day might contribute his little.[96]

During the next decade the pronouncements of the leaders
continued and the next great national report of the condition
of mathematics contained an approval of general mathematics
courses in high school with the following suggestions:

We have already called attention to the fact that, in the earlier
periods of instruction especially, logical principles of organization
are of less importance than psychological and pedagogical principles. In recent years there has developed among many progressive
teachers a very significant movement away from the older rigid
division into "subjects" . . . and toward a rational breaking down
of the barriers separating these subjects, in the interest of an organi-

[94] U. S. Bureau of Education, *Undergraduate Work in Mathematics in Colleges of Liberal Arts and Universities,* Bulletin 1911, No. 7. Washington,
D. C., 1911.
  [95] *Ibid.,* p. 13.
  [96] *Ibid.,* p. 6.

zation of subject matter that will offer a psychologically and pedagogically more effective approach to the study of mathematics.

There has thus developed the movement toward what are variously called "composite," "correlated," "unified," or "general" courses. . . . The movement has gained considerable new impetus by the growth of the junior high school, and there can be little question that the results already achieved by those who are experimenting with the new methods of organization warrant the abandonment of the extreme "water-tight compartment" method of presentation.

The newer method of organization enables the pupil to gain a broad view of the whole field of elementary mathematics early in his high school course. . . . this fact offers a weighty advantage over the older type organization.[97]

In 1929 we find such statements as the following in mathematical literature:

There are various influences which may be said to be bringing about a change in the type of mathematics. . . . Back of these influences are the specialists in education. Perhaps no greater cause of progress in education exists than that of the opinion of a few of the leading specialists in the teaching of particular subjects. . . . Most of the leading educators are expressing opinions favorable to general mathematics.[98]

Dresden called the attention of the joint session of the Association and American Mathematical Society and Section A of A. A. A. S. at Pittsburgh, Pennsylvania, December 31, 1934, to the need for a change in emphasis in elementary college mathematics; and in the following year he again pleaded for the improvement in college teaching in these words:

The need for a shift in emphasis in the teaching of mathematics . . . is one phase of a more general change in orientation which is demanded of all science teaching.[99]

[97] A Report by the National Committee of Mathematics Requirements under the Auspices of the Mathematical Association of America, *The Reorganization of Mathematics in Secondary Education*, p. 13. 1923.

[98] C. McCormick, *Teaching of General Mathematics in Secondary Schools of the United States*, pp. 169-170. Bureau of Publications, Teachers College, Columbia University, New York, 1929.

[99] A. Dresden, "Program for Mathematics." *American Mathematical Monthly*, 42:207, April, 1935.

In 1940 the Joint Commission of the Mathematical Association and the National Council of Teachers of Mathematics acknowledged the attempts to reorganize freshman mathematics in college and they endorsed the general mathematics movement by suggesting two different types of general mathematics courses, one termed "Basic General Course" and the other, "Higher Orientation Course." In commenting upon this type of course, they state:

> Up to the present time, however, colleges and universities have given a very restricted type of mathematics offering for the first two years, though there is a tendency now to design survey courses for the general student. This trend makes it appear likely that the problem of transferring credits from the junior to the senior college in the field of mathematics may be liberalized.
>
> The Commission believes that four-year colleges should give recognition to strong survey courses in junior college, whether or not they themselves offer such work in their first two years.[100]

It is hoped that this retrospective view of the work of the frontier thinkers, the change in the philosophy of education, the growing dissatisfaction with mathematical education, the tendency for curriculum building, the increasing enrollment in educational institutions, the admonitions of the leaders, and the rise of the junior college will serve as a frame of reference from which to view the present status of general mathematics as reported in this study.

---

[100] The Final Report of the Joint Commission of the Mathematical Association of America and The National Council of Teachers of Mathematics, *The Place of Mathematics in Secondary Education,* p. 158. Bureau of Publications, Teachers College, Columbia University, New York, 1940.

CHAPTER IV

OBJECTIVES OF GENERAL MATHEMATICS

OBJECTIVES MUST SATISFY TWO GROUPS

MATHEMATICS students may be roughly divided into two groups, according to their educational objectives. One group consists of those students whose major interest in studying mathematics is to secure a tool subject to be used in a specialized field. This specialized field may be pre-professional, semi-professional, or vocational. But in any case the mathematical knowledge, skills, and techniques are to be used in a particular specialized field. This group may expect "concomitant learnings" of a general educational nature to take place, but these objectives are usually secondary. The particular objectives for the individual courses offered to these students are determined not from the viewpoint of general education but by the requirements of the technical training that is to follow. The courses are designed to prepare the student for further work in mathematics or to give him an understanding of, and the skill to solve, the normal problems that will arise in the profession, semi-profession, or vocation of his choice.

Another group consists of those students whose interests are in professions, vocations, and avocations which do not require great facility in mathematical technique. In this, the great academic group, are those students who wish to acquire an appreciation of the contribution of mathematics to our civilization. They would like to be intelligent citizens—mathematically— but without spending two or more years in mastering techniques and skills which they are confident will be of little value to them. These students would like to possess the mathematical skills and techniques that are worth while to every citizen and to secure those social appreciations of mathematical concepts that are desirable for successful living.

43

OBJECTIVES OF GENERAL EDUCATION

Since the purposes of the latter group are those of general education, it would perhaps be well at this point to recall some of the objectives of general education and to observe the way in which the objectives of general mathematics courses contribute to mathematical education and mathematical education, in turn, contributes to general education.

Cole[1] points out, in a study of 138 articles concerning the objectives of general education during the three periods 1842–76, 1909–21, and 1925–39, that the educational objectives were not static. Moral training and mental discipline represented 42 per cent of the aims in the period 1842–76, 20 per cent in the period 1909–21, and only 11 per cent in 1925–39. On the other hand, citizenship, life's needs, and development of the individual as educational objectives increased from 14 per cent in the period of 1842–76 to 68 per cent in 1925–39. It is also interesting to notice that scholarship as an aim declined between the periods 1909–21 and 1925–39 from 28 to 14 per cent. The study shows that: "In their nature the aims have changed from a concentration upon the purely academic to a concentration upon the personal and social—that is, from the needs of the future scholar to the needs of the future citizen."

The reason for the number of articles concerning the objectives of education is well indicated by the following statement: "There is no dearth of articles concerning the aims and objectives of a college education. These articles contain, however, material that is almost wholly subjective. The aims of education are, in their very nature, matters of opinion; they are not susceptible of objective proof. If they were, there would be no excuse for extended discussions about them. Since personal convictions do not constitute evidence, people can—and probably will—go on arguing about them without ever reaching a tenable conclusion." [2]

---

[1] Luella Cole, *The Background for College Teaching*, p. 33. Farrar and Rinehart, Inc., New York, 1940.
[2] *Ibid.*, p. 15.

In spite of the difference of opinions concerning the detailed aims of education the objectives seem to center around the aim of developing the desirable potential abilities of the individual to the full and providing experiences that will condition him to become a citizen of maximum worth in a democratic society. This dual aim of education is indicated in the majority of the discussions of educational objectives, of which the following is typical: "The purpose of general education is to provide rich and significant experiences in the major aspects of living, so directed as to promote the fullest possible realization of personal potentialities, and the most effective participation in a Democratic Society." [3]

OBJECTIVES OF MATHEMATICS AS SUGGESTED BY COMMITTEES

If mathematics is to be a part of general education, then its aims must supplement and be a part of the objectives of general education. With this intention for a guide, two purposes are beginning to make themselves prominent in mathematics education. These purposes are indicated by the Final Report of the Joint Commission of the Mathematics Association of America and the National Council of Teachers of Mathematics as follows: ". . . objectives may be regarded as having either a factual and impersonal aspect or a personal, psychological bearing. Thus, when we study a given domain in a purely scientific way, irrespective of the learner's personal reactions, we are mainly interested in facts, skills, organized knowledge, accurate concepts, and the like. If, on the other hand, we scrutinize the way in which the pupil behaves in a given situation, or his modes of reaction, we are led to such categories as habits of work or study, attitudes, interests, insight, modes of thinking, types of appreciation, creativeness, and the like. A clear recognition of these two essentially different yet complementary types of ob-

---

[3] Progressive Education Association, *Mathematics in General Education*, p. 43. D. Appleton-Century Company, New York, 1940.

jectives is one of the achievements of recent educational theory." [4]

The above report discusses the objectives of mathematics in general education in terms of mathematical information, such as the concepts, principles and skills of mathematics and also in terms of desirable attitudes and mathematical appreciation. Thus these two objectives, which may not be independent of each other, are emphasized by one of the latest committee reports in mathematics education.

In the report of the Committee on the Function of Mathematics in Terminal Education, for the Commission on Secondary School Curriculum[5] "the role of mathematics in achieving the purpose of general education" is described at length. The objectives as proposed by the Committee fall into two main categories: One, "Role of mathematics in meeting the needs of the students," and, two, "Role of mathematics in developing personal characteristics essential to democratic living," such as social sensitivity, aesthetic appreciation, self-direction, creativeness, etc.[6]

In like manner, in the Report of the National Commission on Cooperative Curriculum Planning,[7] "the contributions of mathematics to general education" may be classified under two headings. First, it is possible for a student to profit by the study of mathematics because of its value in problem solving as a tool subject and mode of thinking. Second, mathematics may make a contribution in the development of interests, attitudes, and appreciations.

Thus the objectives of mathematics education as indicated by reports of committees in the field contribute to the two general

[4] Joint Commission of the Mathematical Association of America and the National Council of Teachers of Mathematics, *The Place of Mathematics in Secondary Education*, pp. 21, 22. The Final Report. Bureau of Publications, Teachers College, Columbia University, New York, 1940.

[5] Progressive Education Association, *Mathematics in General Education*, p. 43. D. Appleton-Century Company, New York, 1940.

[6] *Ibid.*, pp. 43-53.

[7] National Commission on Cooperative Curriculum Planning, *The Subject Fields in General Education*, pp. 136-147. D. Appleton-Century Company, New York, 1941.

educational objectives—namely, to develop the potentialities of the student for individual achievement, and to condition him for successful citizenship.

OBJECTIVES OF GENERAL MATHEMATICS AS SUGGESTED BY THE AUTHORS

The questions might be raised, "What are the educational objectives of the *general* mathematics courses?" "Are the objectives of *general* mathematics courses in harmony with the objectives of general education?" One way of answering these questions is by examining the educational objectives given by the authors of the general mathematics textbooks. A careful examination was made of the objectives given by more than fifty authors of general mathematics textbooks.

Preparatory Objectives

The objectives of the general mathematics course are definitely stated in the writing of many authors of textbooks. For example, Woods and Bailey, authors of one of the first textbooks in general mathematics in the United States, clearly state: "In the preparation of the text the needs of a student who desires to use mathematics as a tool in engineering and scientific work have been primarily considered, but it is believed that the course is also adapted to the student who studies mathematics for its own sake." [8]

Young and Morgan, authors of a textbook published a decade later, say: "This book aims to present a course suitable for students in the first year of our colleges, universities, and technical schools." [9]

The following statements expressing the opinions of authors concerning the objectives of general mathematics are typical of statements of purposes in a certain group of general mathematics textbooks:

[8] F. S. Woods and F. H. Bailey, *A Course in Mathematics,* v. Ginn and Company, Boston, 1907.
[9] J. W. Young and F. M. Morgan, *Elementary Mathematical Analysis,* v. The Macmillan Company, New York, 1917. Quoted by permission of the publisher.

. . . it has been the aim to give the fundamental truths of elementary analysis as much prominence as seems possible in a working course for freshmen.[10]

This text presents a course in elementary mathematics adapted to the needs of students in the freshman year of an ordinary college or technical school course, and of students in the first year of a junior college. The material of the text includes the essential and vital features of the work commonly covered in the past in separate courses in college algebra, trigonometry, and analytical geometry.[11]

The fundamental idea of the development is to emphasize the fact that mathematics cannot be artificially divided into compartments with separate labels, as we have been in the habit of doing, and to show the essential unity and harmony and interplay between the two great fields into which mathematics may properly be divided: viz., analysis and geometry.[12]

A further fundamental feature of this work is the insistence upon illustrations drawn from fields with which the ordinary student has real experience. The authors believe that an illustration taken from life adds to the cultural value of the course in mathematics in which this illustration is discussed. Mathematics is essentially a mental discipline, but it is also a powerful tool of science, playing a wonderful part in the development of civilization. Both of these facts are continually emphasized in this text and from different points of approach.[13]

It aims to combine the work which is commonly covered in separate courses in college algebra and in analytical geometry.[14]

The aim in selecting material and in the manner of presenting it has been to acquaint the student with as much of the content of mathematics as possible and to train him in the facile use of mathematics as a tool.[15]

This book is written for the purpose of giving students who are beginning their college work, . . . a general view of the meaning

[10] C. S. Slichter, *Elementary Mathematical Analysis*, vii. The McGraw-Hill Book Company, Inc., New York, 1917.

[11] L. C. Karpinski, H. Benedict, and J. W. Calhoun, *Unified Mathematics*, iii. D. C. Heath and Company, New York, 1918.

[12] *Ibid.*

[13] *Ibid.*

[14] E. R. Breslich, *Correlated Mathematics for Junior Colleges*, ix. The University of Chicago Press, Chicago, 1928.

[15] J. E. Rowe, *Introductory Mathematics*, iii. Prentice-Hall, Inc., New York, 1927.

of the mathematics that immediately follow the rudiments of algebra and geometry.[16]

It is the hope of the authors that this book will thus meet the need in many colleges for a presentation of concrete mathematics that will stimulate students to continue their interest in a science which touches so many phases of human activity.[17]

The order in which the topics are presented in this book is such that in a one year course, three times a week, a student who has had only elementary algebra and plane geometry in high school may be led gradually from familiar ideas of algebra to the solution of the nth degree equation, then through the basic ideas of trigonometry and the analytic geometry of the straight line to the differential and integral calculus.[18]

The introduction contains a collection of review topics for reference from elementary algebra and plane geometry. Chapters I to IX will be found to be a one term course in Advanced Algebra and Trigonometry; Chapters X to XVII present an elementary one term course in Analytic Geometry and Calculus.[19]

Chapters III to XI inclusive form a reasonably complete course in algebra . . . Chapters XII and XIII are devoted entirely to trigonometry and, together with those in algebra, furnish the material for five semester-hours of instruction. The book is thus adapted to courses which aim at the preparation in one semester of a student for analytic geometry and calculus.[20]

As the study of the derivative progresses, additional mathematical processes are developed and the student is enabled to acquire a fair understanding of the branches of mathematics usually included under the titles solid geometry, trigonometry, and analytic geometry, differential calculus and integral calculus, not as separate subjects of study but as a unified system of thought, namely, mathematical analysis.[21]

[16] G. W. Mullins and D. E. Smith, *Freshman Mathematics,* iii. Ginn and Company, Boston, 1927.

[17] *Ibid.,* iv.

[18] C. H. Helliwell, A. Tilley, and H. F. Wahlert, *Fundamentals of College Mathematics,* v. The Macmillan Company, New York, 1935. Quoted by permission of the publisher.

[19] *Ibid.*

[20] H. T. Davis, *A Course in General Mathematics,* ix. The Principia Press, Bloomington, Indiana, 1935.

[21] M. Philip, *Mathematical Analysis,* viii. Longmans, Green and Company, New York, 1936. Quoted by permission of the publisher.

The present text is the result of an effort to reorganize the material usually given in freshman college mathematics. The mathematical equipment which the average student brings to college is frequently limited, yet each year the boundary line of knowledge in mathematics and its allied fields is pressed forward along many fronts into newer and larger domains. If the student is to be provided with a tool that will permit him to take part in this development, or if he is to be given a cultural background which will enable him to appreciate its significance, it is necessary that all duplication, all waste motion be eliminated, and that the essentials of mathematical procedure be presented in as simple, as direct, and as unambiguous a manner as possible.

In the present text algebra is treated as a tool subject—algebraic concepts are introduced and developed . . . for the study of trigonometry and analytic geometry. Fractions, the simplification of radicals, linear equations, the properties of quadratic equations and determinants are studied for the purpose of their immediate use. This use serves to crystallize them in the mind of the student.[22]

The author has made a definite bid for interest, independence of thought and initiative, and has tried to develop an understanding of the fundamental principles of the whole subject of introductory mathematics and to correlate it with other fields of science. . . . One of the author's main objectives has been to write a text requiring as little memorization as possible, but rather, to develop an understanding of the fundamental and underlying principles of the subject. Another objective has been to stimulate independence of thought and to encourage initiative.[23]

The underlying theme of Introductory Mathematical Analysis is the concept of function. The mathematical concepts and processes traditionally presented in separate courses, as college algebra, trigonometry, and analytic geometry, are herein presented as separate manifestations of the basic concept of function. A comprehensive treatment of functionality necessitates the utilization of the processes of differentiation and integration. These processes are introduced where they may be employed to advantage in the rationalization of the particular function under consideration.[24]

[22] F. E. Johnston, *Introductory College Mathematics*, iii. Farrar and Rinehart, Inc., New York, 1940.

[23] V. H. Wells, *First Year College Mathematics*, iii. D. Van Nostrand Company, Inc., New York, 1937.

[24] J. S. Georges and J. M. Kinney, *Introductory Mathematical Analysis*, v. The Macmillan Company, New York, 1938. Quoted by permission of the publisher.

It is not intended for a survey course; neither is it expected to give a smattering of the first two years of college mathematics. Although it is an attempt to meet the demands of those curricula in which mathematics enters in a utilitarian role, the author believes that any group of college students who need the subject matter of college algebra, plane trigonometry, and plane and solid analytic geometry can use this book to advantage.[25]

The objectives suggested in the above quotations all center around one general aim. That aim is to prepare students for further work in mathematics or to use mathematics as a tool in a specialized field, such as science, engineering, and semi-technical fields. Other objectives are suggested but only as a secondary and never as a primary aim of the course.

Cultural Objectives

Another group of textbooks, nearly as large and increasing in number, were written by authors who claim the following objectives:

A very large group of freshmen taking mathematics do not continue the study of this subject in the following years, and the needs of these students have received primary consideration.[26]

The choice and arrangement of topics have been definitely influenced by the desire for simplicity, continuity, and perspective. . . . In the beginning the aim was to construct a single (quarter or semester) course which would give to the students a rather precise idea as to the nature of the fundamental notions of elementary mathematics and their relation to his everyday life. The first volume is offered as meeting this aim.[27]

In the preparation of this text we have had as our main objective a course which would furnish the ordinary student in the junior college those facts and processes of mathematics which are necessary for the proper understanding of elementary required courses, or for the intelligent reading of newspapers and magazines of today; in fact, this course seems to furnish an almost irreducible minimum of

[25] Reprinted by permission from *A First Year College Mathematics*, v, by H. J. Miles. Published by John Wiley and Sons, Inc., New York, 1941.
[26] A. S. Gale and C. W. Watkeys, *Elementary Functions and Applications*, iv. Henry Holt and Co., New York, 1920. Quoted by permission of the publisher.
[27] M. I. Logsdon, *Elementary Mathematical Analysis*, Volume I, v. McGraw-Hill Book Company, Inc., New York, 1932.

mathematics for those who plan to complete the junior college.[28]

The feeling persisted that for a person who is willing to make a serious effort, there should be a way of acquiring sufficient comprehension of the nature of mathematics to enable him to understand without having a large technical equipment, the place this subject occupies in the world of thought and that such understanding should form a part of the equipment of an educated person.[29]

. . . to give a reader who has but little knowledge of the technique of mathematics, an insight into the character of at least some of the important questions with which mathematics is concerned, to acquaint him with some of its methods, to lead him to recognize its intimate relation to human experience and to bring him to an appreciation of its unique beauty. These aims in turn determined to a considerable extent the contents of the book and the method of presentation.[30]

It is the belief of the authors that most students who do not specialize in mathematics leave its study without having acquired any real understanding of the character of the subject or of its relations to the sciences, the arts, philosophy, and to knowledge in general. Too often they have been taught little more than a variety of techniques in special branches of mathematics and thus have acquired a narrow, distorted, and hence incorrect view of mathematics. As a result, far too many intelligent students are "soured" for life as far as mathematics is concerned. It is only too well known that many capable students try to avoid college mathematics because of the suspicion that the material is boring and to a large extent useless to them because of its highly technical nature. Even students who specialize in mathematics usually acquire no proper understanding of the nature of the subject or of its relations to the other fields of knowledge until quite late in their college years. This book attempts to remedy this unfortunate situation by showing that mathematics has much more to offer serious-minded college students than mere training in memorizing formulas and manipulating symbols. The principal objectives of the book are the following:

---

[28] R. P. Stephens, D. F. Barrow, and W. S. Beckwith, *Freshman Mathematics*, Preface. Division of Publication, The University of Georgia, Athens, Georgia, 1936.

[29] A. Dresden, *Invitation to Mathematics*, v. Henry Holt and Company, New York, 1936. Quoted by permission of the publisher.

[30] A. Dresden, *Invitation to Mathematics*, vi. Henry Holt and Company, New York, 1936. Quoted by permission of the publisher.

1. To show how many of the fundamental ideas of mathematics have their sources in physical experience.

2. To show how, from these ideas, mathematics builds broad logical theories which have wide applications in the physical, biological, and social sciences, the arts, and philosophy.

3. To show that mathematics is not merely a collection of methods useful in the sciences, but a vast unified system of reasoning which possesses many of the characteristics of a fine art.

4. To acquaint the student with the logical structure of a mathematical system and thus provide him with a standard of exact reasoning which should help him to achieve a more critical attitude toward conclusions arrived at in other fields.

5. To show that science and philosophy are indebted to mathematics for many precise concepts, such as velocity, motion, and infinity.

6. To open the student's mind to the fact that the development of mathematics from ancient to modern times has been an important factor in the development of civilization.[31]

For most of the sciences the veil of mystery is gradually being torn asunder. Mathematics, in large measure, remains unrevealed. What most popular books on mathematics have tried to do is either to discuss it philosophically, or to make clear the stuff once learned and already forgotten. In this respect our purpose in writing has been somewhat different. . . . It has been our aim . . . to show by its very diversity something of the character of mathematics, of its bold, untrammeled spirit, of how, as both an art and a science, it has continued to lead the creative faculties beyond even imagination and intuition.[32]

In writing this book we had in mind two distinct but, we trust, compatible goals. The first was the production of a college textbook which would provide enough of the conventional subject matter to meet practical credit-transfer requirements. The second was the goal of high-lighting for the non-specialists the interest that is inherent in mathematics itself and of fostering an appreciation of its place in modern life. Our aim, in short, is to answer in the text itself that perennial and petulant student query: "What's the good of all this?"

The book provides a one-year course for those who will theoretically pursue the subject no farther, but among whom there may

[31] H. R. Cooley, D. Gans, M. Kline, and H. E. Wahlert, *Introduction to Mathematics*, iii, iv. Houghton Mifflin Company, Boston, 1937.

[32] E. Kasner and J. Newman, *Mathematics and the Imagination*, xiv. Simon and Schuster, New York, 1940.

possibly be salvaged a few devoted and surprised lifetime addicts.[33]

While one who has mastered our book will not be an accomplished mathematician (how many sophomores are?), he will have done about as much plain thinking as is expected of most freshmen. . . . We are writing, let us repeat, for the great battalion of those who fear the subject—not with the idea of removing their difficulties, but rather with the hope of adding interest and pleasure to their work.[34]

In recent years many college students are regarding some training in mathematics as an essential part of a general education, but they do not expect to do extensive work in the field. In fact, a brief survey has indicated that in many colleges and universities almost half of the students who study mathematics take only a single course. Thus there has been considerable discussion among educators in regard to an introductory course in mathematics for such students. Many who have considered the problem believe that no single one of the traditional college courses is suitable; the author agrees with this point of view. This manuscript, then, contains the basic argument of a course which has been developed at the University . . . during the past eight years to accommodate this particular group of students.[35]

What should be given to students of the arts and social sciences, in what is usually their last year of mathematics? The customary freshman course with emphasis on further memorized and regurgitated techniques seems to be far from the best way to use this educational opportunity. A solution of this problem must rest, in my opinion, on two hypotheses, of whose truth I am firmly convinced, namely, that mathematics has something of value to offer these students, and that the students are not necessarily unintelligent whether or not they are proficient in the routine manipulations of their high school algebra. This book is intended primarily to provide a sound, suitable and elastic course for the class of students mentioned; however, considerations of the kind discussed here would also be beneficial for students of the sciences who are commonly and naively expected to acquire an understanding of fundamental concepts by osmosis.

The principal objectives of this book are to give the student:

[33] R. S. Underwood and F. W. Sparks, *Living Mathematics*, v. McGraw-Hill Book Company, Inc., New York, 1940.

[34] *Ibid.*, vi.

[35] C. V. Newsom, *An Introduction to Mathematics*, Introduction. The University of New Mexico Press, Albuquerque, 1940.

(1) An appreciation of the natural origin and evolutionary growth of the basic mathematics ideas from antiquity to the present; (2) A critical logical attitude, and a wholesome respect for correct reasoning, precise definitions, and clear grasp of underlying assumptions; (3) An understanding of the role of mathematics as one of the major branches of human endeavor, and its relations with other branches of the accumulated wisdom of the human race; (4) A discussion of some of the simple important problems of pure mathematics and its applications, including some which often come to the attention of the educated layman and cause him needless confusion; (5) An understanding of the nature and practical importance of postulational thinking.[36]

The material . . . is designed for a freshman course for non-science students.[37]

To relieve mathematics of the mystery surrounding it because its aims and methods are so seldom appreciated; to expose its utility, philosophy, and beauty; to accomplish these ends with elementary but significant material, available to all who, regardless of previous mathematical experience, are willing to cultivate the ability to reason; to present that material in interesting and stimulating style—these are the purposes of this book. . . . Clearly, the aim of our exploration cannot be the cultivation of mathematics as a tool. We cannot insist on technical ability to begin with nor as an essential result of even a true comprehension of the essays. Emphasis must be on ideas rather than on techniques.[38]

Attention should be called to the cultural mathematics textbooks that are used as a survey course for all students. These textbooks are designed for a one-semester or quarter survey course to precede the work in traditional mathematics, and, as it has been pointed out by some of the authors of these textbooks, they may be used by both the terminal and the preparatory student; but it is not their purpose to replace the mathematics courses as the cultural-preparatory textbooks attempt to do.

[36] M. Richardson, *Fundamentals of Mathematics*, v, vi. The Macmillan Company, New York, 1941. Quoted by permission of the publisher.

[37] J. H. Zant and A. H. Diamond, *Elementary Mathematical Concepts*, Foreword. Burgess Publishing Company, Minneapolis, Minnesota, 1941.

[38] Reprinted by permission from *To Discover Mathematics*, vii, by G. M. Merriman, published by John Wiley and Sons, Inc., New York, 1942.

. . . It intends rather to serve as a guide to the subject. It is a book not primarily of mathematics but about mathematics. Its purpose is to give to one who has studied the traditional algebra and geometry of the high school, be he general reader or student, a broad outlook over the entire field of elementary mathematics through the calculus, thus correcting the perspective of one who has failed to see the forest because of the trees.[39]

The attempt to give in a few lectures a vivid picture of the historical development of the mathematics of classical times with a description of the types of problems which led to the growth of elementary concepts of arithmetic, algebra, geometry and trigonometry, and to give something of the purport and processes of the modern subjects, analytical geometry and the calculus, to the end that the student may obtain fairly definite ideas of their meanings and uses in modern life and of their relations to the various fields of the physical sciences, has been rendered more difficult than pleasant by the lack of satisfactory references for extensive reading; and it is to meet that need that this book has been written. . . . It does not take the place of any one or more texts in the standard courses in college mathematics, but its sponsors believe that it will prove to be of use along the following lines:

1. To provide the mathematics for general physical science courses, as at the University of Chicago.

2. To serve as a text for one-hour or two-hour orientation course in college, junior college, or senior high school.

3. To serve as a reading reference for first-year and second-year mathematical courses in college or junior college.

4. To serve as a supplementary text for courses in the teaching of mathematics in normal schools and teachers colleges.

5. To serve as an eye-opener for the adult who knows no mathematics beyond elementary algebra and geometry but who has a healthy curiosity concerning the science whose development has made possible this age of the machine.[40]

It is believed that this new Survey Course is one which college freshmen in general could judiciously be advised to take—whether as a final course for its informational value concerning the role of mathematics in the world, or as a means of determining whether the student wishes to pursue the study of mathematics further, or as a preliminary reconnaissance on the part of those who already know

[39] R. E. Bruce, *A Survey of Elementary Mathematics*, p. 7. Boston University Book Stores, Boston, Massachusetts, 1936.

[40] M. I. Logsdon, *A Mathematician Explains*, vi. The University of Chicago Press, Chicago, 1935.

their inner urge to specialize in this field, but who will profit greatly by a preliminary view of its general character and significance.[41]

Textbooks representative of these objectives are not numerous,[42] but they represent an attempt to transmit some of the concepts of mathematics as a social heritage and at the same time serve as an exploratory course for college freshmen.

The above objectives indicate a course in mathematics whose primary purpose is not to prepare students for professions or vocations in which they will use mathematics as a tool subject. Nor is it necessarily a course of the same content as the traditional mathematics with less emphasis on skill and techniques. Rather, it is a course which emphasizes an understanding and an appreciation of the larger significant concepts of mathematics and the relationship of its contribution to civilization. The ideas, as well as the skills, arc beneficial for the successful individual as a citizen and the mathematical concepts should be a part of his social heritage.

Cultural-Preparatory Objectives

It might appear, since the students seem to divide themselves into two groups in mathematics—the preparatory and the terminal groups—that the general mathematics courses would be divided on the bases of these two objectives. The objectives just quoted from authors seem to fall into these two classes. But other authors have suggested that general mathematics might satisfy both objectives with the same or nearly the same content material. For example:

This book aims to give the student an idea of the kinds of problems met with and the tools used in the chief courses in mathematics of college grade. It may take the place of fuller single courses in trigonometry, college algebra, and solid geometry in the last year of the high school or the first year of college. While it contains enough material to prepare the students for any of the mathematics associated with the sophomore year of American Colleges, it is possible by omitting chapters 3, 7, 12, 13, 14, 23, and 24 to use it as a

[41] N. J. Lennes, *A Survey Course in Mathematics*, xi. Harper and Brothers, New York, 1926.
[42] See Appendix for list of mathematics textbooks.

textbook in a course for those who are not to pursue further mathematical studies and desire a broad knowledge of the subject matter and principles of mathematical analysis without spending time on the development of a technical facility in algebraic manipulations.[43]

The students for whom it was written will fall naturally into two classes: those who will terminate their study of mathematics with the usually required freshman course, and those who will continue the study of either pure or applied mathematics. The authors have endeavored to keep in mind the needs of both of these groups.[44]

Students of college mathematics may be roughly classified into three groups: (*a*) those who take only one year of mathematics; (*b*) those who take mathematics as a specific service course for the physical and natural sciences, engineering, statistics, etc.; (*c*) those who take mathematics as their major field of work and who expect either to teach this subject or to specialize in this line of work. This text is an outgrowth of an effort to meet the needs of all three groups.[45]

Under the traditional plan of studying trigonometry, college algebra, analytic geometry, and calculus separately, a student can form no conception of the character and possibilities of modern mathematics, nor of the relations of its several branches as parts of a unified whole until he has taken several successive courses. Nor can he, early enough, get the elementary working knowledge of mathematical analysis, *including integral calculus,* which is rapidly becoming indispensable for students of the natural and social sciences. Moreover, he must deal with complicated technique in each introductory course; and must study many topics apart from their uses in other subjects, thus missing their full significance and gaining little facility in drawing upon one subject for help in another. To avoid these disadvantages of the separate-subject plan the unified course presented here has evolved. This enables even those students who can take only one semester's work to get some idea of differential and integral calculus, trigonometry, and logarithms.[46]

---

[43] W. R. Ransom, *Freshman Mathematics,* iii. Longmans, Green and Company, New York, 1925. Quoted by permission of the publisher.

[44] J. W. Lasley and E. T. Browne, *Introductory Mathematics,* v. McGraw-Hill Book Company, Inc., New York, 1933.

[45] W. E. Milne and D. R. Davis, *Introductory College Mathematics,* iii. Ginn and Company, Boston, 1935.

[46] F. L. Griffin, *Introduction to Mathematical Analysis,* iii. Houghton Mifflin Company, Boston, 1936.

The general aim is still to present these subjects so that the student may have a real understanding of the fundamental principles and processes involved and of the values of these subjects vocationally and culturally. It is hoped that the text will give the student an adequate foundation of mathematics, irrespective of his educational objectives.[47]

The text is, in consequence, especially suitable for use in colleges of engineering and for the training of students preparing to teach mathematics in secondary schools and of any other students who expect to study calculus. But because of the great interest and utility of the material and the unity of its presentation, it also provides a very satisfactory terminal course in mathematics.[48]

In mapping out a course in mathematics to be used by students in their first year of college, it is necessary to keep in mind the benefits which the students themselves hope to derive from such a course. To many of them it is an entrance into the broad field of mathematics and science, and they desire to obtain a sound basis for future work. Others, expecting to pursue non-scientific studies, wish to learn in a year's time something of the nature and scope of the subject. The authors hope that this text will meet the needs of both classes of students.[49]

Until a few years back, freshman mathematics consisted of separate courses in algebra and trigonometry. In a few of the most advanced universities analytics and the calculus was included. Then came the innovation of a unified course, in which all four of these subjects were intermingled at a distinct sacrifice in many instances of algebra and trigonometry. Recently there seems to have developed an unfavorable reaction by some to the usual unified course. This dissatisfaction is due in part to the fact that students with either average or poor preparation find the work extremely difficult. As a result, many complete the course without learning enough algebra and trigonometry to enable them to pursue successfully further courses in mathematics or the sciences. On the other hand, students who have been well prepared in algebra find excessive duplication later if they take separate courses in analytics and the calculus.

[47] H. I. Slobin and W. E. Wilbur, *Freshman Mathematics*, v. Farrar and Rinehart, Inc., New York, 1938.

[48] R. W. Brink, *A First Year of College Mathematics*, v, vi. D. Appleton-Century Company, Inc., New York, 1937.

[49] M. A. Hill, Jr., and J. B. Linker, *Introduction to College Mathematics*, iii. Henry Holt and Company, New York, 1938. Quoted by permission of the publisher.

We have tried to use the best of both plans in this book, making it consist essentially of algebra and trigonometry, treated separately, with a very slight introduction to analytics and the calculus by way of application of algebra. . . . Thus we try to make this a basis for a cultural course in mathematics for those students who do not wish to go further, and at the same time to give as thorough a preparation as possible for those who wish to continue the subject.[50]

The purpose of this text is to present in one volume the essentials of college algebra and plane trigonometry, and an introduction to plane analytic geometry and calculus. . . . The authors believe that college students who take only one year of mathematics should acquire a knowledge of the essentials of several of the traditional subjects.[51]

The text is designed to meet the needs of two different groups of students: (1) those taking only one course in collegiate mathematics, for whom the text serves as a survey and (2) those preparing for further work in mathematics, for whom the text serves as an introduction. The authors believe that the latter group should follow this course with one in calculus; repetition of the elementary concepts should improve the student's understanding of the subject.[52]

The text is suitable for students in pre-engineering courses, those preparing to teach mathematics, and those who wish only one year of college mathematics in preparation for other professions.[53]

A study of these objectives indicates that some authors of general mathematics textbooks believe that it is possible to design a course in general mathematics that will give adequate preparation for the student who takes further work in mathematics and also will lend itself readily to the needs of the large academic group of terminal mathematics students. This point of view is open to serious question and will be discussed in more detail as this investigation continues.

[50] E. L. Mackie and V. A. Hoyle, *Elementary College Mathematics,* iii. Ginn and Company, Boston, 1940.

[51] W. W. Elliott and E. R. C. Miles, *College Mathematics—A First Course,* v. Prentice-Hall, Inc., New York, 1940.

[52] C. W. Munshower and J. F. Wardwell, *Basic College Mathematics,* v. Henry Holt and Company, New York, 1942. Quoted by permission of the publisher.

[53] C. C. Richtmyer and J. W. Foust, *First Year College Mathematics,* v. F. S. Crofts and Company, New York, 1942.

Summary of Authors' Objectives

Thus, a survey of the objectives of general mathematics as given by the authors of more than fifty general mathematics textbooks indicates that the aims of general mathematics courses fall into three categories: (1) To prepare students for a profession, semi-profession, or vocation in which mathematics is useful as a tool and emphasis is placed on facility in mathematical manipulation as well as on understanding of the concepts involved. (2) To prepare students to be intelligent citizens, mathematically—to transmit those concepts and skills that are the necessary equipment of a desirable citizen and to emphasize the understanding of mathematical concepts, the relation of mathematics to other great fields of learning with little emphasis on the manipulative aspects in problem solving. (3) To attain both the above objectives by meeting the needs of the large academic terminal mathematics group and also to furnish an adequate preparation for the minority who wish to pursue further courses in mathematics.

OBJECTIVES OF GENERAL MATHEMATICS AS SUGGESTED BY THE TEACHERS

The question "What are the objectives of general mathematics?" has been answered by the authors of the textbooks and committees of specialists in the field, but how would this question be answered by the many teachers who are using these textbooks? Are their objectives for general mathematics in harmony with those of the authors whose text they are using?

Teachers' Objectives as Indicated by a Questionnaire

A study of the replies to a questionnaire sent to more than five hundred colleges that listed in their catalogues a course whose description indicated that it might be a *general* mathematics course showed that in many cases the primary objective of the school offering the course was not the major objective of the author of the textbook. Of those institutions using a textbook whose objective according to the author was preparatory, nearly one half indicated on the questionnaire other objectives in place of, or in addition to, the author's primary objective. In fact, more than one fourth of the reporting schools using this type of textbook did not indicate "preparation for further work in mathematics" as one of the objectives. From the responses to item 9[54] and the comments which appeared freely on the questionnaires, it appears that the objective of the instructors offering preparatory general mathematics is not in all cases in harmony with that of the author of the textbook used.

Of the instructors using a cultural-preparatory textbook, three fourths indicated a single primary objective. Sixty-four per cent of these taught a course preparatory for further study in mathematics. Eighteen per cent offered a terminal course primarily for practical mathematical education. And eighteen per cent gave a terminal course primarily for a cultural education.

Teachers' Objectives as Indicated by Interviews

This disagreement between the objectives of the authors and the objectives of users of the textbook was substantiated by interviews with many teachers. Nearly a third of those interviewed proposed objectives other than the author's, and when an author's objective was suggested they conceded that it might be a secondary one but not the primary one for their students.

Since the instructors and the authors do not seem to agree on the purpose of the general mathematics course, perhaps it is proper to raise the following questions: What are the proposals to meet these objectives? How do the authors propose to meet the objectives that they have suggested? How do the instructors attempt to meet their aims? The answer to these questions is the subject of the following chapter.

[54] See Questionnaire A in Appendix B.

CHAPTER V

PROVISIONS FOR MEETING OBJECTIVES OF GENERAL MATHEMATICS

PROVISIONS AS INDICATED BY THE AUTHORS OF PREPARATORY TEXTBOOKS

S OME of the ways in which the authors of textbooks are attempting to provide a satisfactory preparation for further study in mathematics and those sciences requiring mathematical facility are as follows: increased emphasis on understanding in addition to skills and techniques necessary in computing; greater stress upon the importance of the function concept; a breaking down of the compartmental lines; a change in the order of the presentation of the topics; and a variation in the thought-provoking exercises, applications, and illustrations given. With the reorganization of the content and the inclusion of non-traditional mathematical material, it might be expected that the authors would indicate requirements both in prerequisites and in time to complete the general mathematics course different from the requirements of the traditional course. But as will be shown in the discussion which follows, this is not the case.

Emphasis on Meaning

There is a desire and an attempt on the part of many of the authors to emphasize the meaning as well as the ability to perform mechanically mathematical operations.

Typical of this attitude is the statement of an author of a well-known general mathematics textbook, who says: "The methods used and the subject matter included in first-year courses in mathematics in American institutions of higher learning are undergoing marked changes. It is no longer possible to interest a student by training him in the techniques of mathematics alone, nor is it desirable." [1]

Function Concepts

Among the ideas for the improvement of mathematical education that have received major attention by many of the authors of the preparatory general mathematics textbooks is the function concept. The importance of this concept in the opinion of textbook writers is shown by their statements:

The ideas explained above are developed in accordance with a two-fold plan, as follows: First, the plan is to group the material of elementary analysis about the consideration of the three fundamental functions: 1. The Power Function $y = ax^n$ . . . 2. The Simple Periodic Function $y = a \sin mx$ . . . 3. The Exponential Function . . . Second, the plan is to enlarge the elementary functions by the development of the fundamental transformations applicable to these and other functions. . . . The emphasis of the book is intended to be upon the notion of functionality. [2]

The concepts of a function and of its derivative are introduced at the outset in the solution of a simple problem. As a result, algebra and geometry lose their separate identities and are found to be merely different aspects of one mode of thought. [3]

The concept of function is used as a unifying principle about which the various concepts and processes of the course are correlated. The algebraic and geometric aspects are treated jointly, each enhancing the interpretation of function inherent in the other. [4]

Disregard of Compartmental Lines and Introduction of Calculus

The emphasis upon the function concept as an important unifying principle in the organization of the content of a gen-

[1] V. H. Wells, *First Year College Mathematics,* iii. D. Van Nostrand Company, Inc., New York, 1937.

[2] C. S. Slichter, *Elementary Mathematical Analysis,* vii. McGraw-Hill Book Company, Inc., New York, 1918.

[3] M. Philip, *Mathematical Analysis,* vii. Longmans, Green and Company, New York, 1936. Quoted by permission of the publisher.

[4] J. S. Georges and J. M. Kinney, *Introductory Mathematical Analysis,* v. The Macmillan Company, New York, 1938. Quoted by permission of the publisher.

eral mathematics course for students preparing for further study in mathematics has encouraged the breaking down of the traditional compartmental lines.

In arranging the material, however, the traditional division of mathematics into distinct subjects is disregarded, and the principles of each subject are introduced as needed and the subjects developed together. The objects are to give the student a better grasp of mathematics as a whole, and of the interdependence of its various parts, and to accustom him to use, in later applications, the method best adapted to the problems in hand.[5]

By means of it [the calculus] the study of other subjects is materially aided and the continued application of the calculus ideas and methods to all sorts of algebraic and analytic problems gives the student a better knowledge of the calculus than by any other method. Furthermore, it is believed that the one-year student will profit more by a knowledge of the calculus concepts and the ability to apply them than he will by complete courses in college algebra and analytic geometry alone.[6]

An author of one of the first textbooks published in the field of general mathematics says, after suggesting that certain obsolete material should be removed from all freshman textbooks:

At the same time a decided advantage is gained in the introduction of the principles of analytic geometry and calculus earlier than is usual. In this way these subjects are studied longer than is otherwise possible, thus leading to greater familiarity with their methods and greater freedom and skill in their applications.[7]

That not all authors have agreed with this view is indicated by the following remark:

No attempt has been made to introduce the terminology of the calculus as it is found that there is ample material in the more elementary field which should be covered before the student embarks upon what may properly be called higher mathematics.[8]

[5] F. S. Woods and F. H. Bailey, *A Course in Mathematics,* iii Ginn and Company, Boston, 1907.
[6] V. H. Wells, *First Year College Mathematics,* iii, iv. D. Van Nostrand Company, Inc., New York, 1937.
[7] F. S. Woods and F. H. Bailey, *A Course in Mathematics,* iii. Ginn and Company, Boston, 1907.
[8] L. C. Karpinski, H. Y. Benedict, and J. W. Calhoun. *Unified Mathematics,* iv. D. C. Heath and Company, New York, 1918.

Some of the authors have advocated organizing the content around functionality; others emphasize the importance of the calculus as a central theme, while still others say:

> . . . the fundamental principle of the course is to associate closely mathematical topics which are naturally related to each other. This combination makes it possible better to motivate each topic, to show the student more clearly the meaning of the subject by means of geometrical representation, and to develop in a natural way the important concept of functional correspondence.[9]

In any case the reverence for the traditional air-tight compartments of mathematics is disappearing.

Change in Order of Topics

There are slight changes in the sequence of the topics, such as those suggested below:

> Some features of the book are the following:
> 1. An initial chapter on arithmetic . . .
> 2. The trinometer and trinometry . . .
> 3. For the first time in any text, we have general method of solving the quadratic, the cubic, and the biquadratic equations; that is, the same method may be used to solve all three equations.[10]

The author believes that the subject of trigonometry does not lend itself to a correlation with the subjects of algebra, analytic geometry and elementary calculus, but that the latter subjects are closely related and are readily studied simultaneously.[11]

It is difficult to understand why it is customary to introduce the trigonometric functions to students seventeen or eighteen years of age by means of the restricted definitions applicable only to the right triangle. Actual test shows that such rudimentary methods are wasteful of time and actually confirm the student in narrowness of view and in lack of scientific imagination.[12]

---

[9] E. R. Breslich, *Correlated Mathematics for Junior Colleges,* ix. The University of Chicago Press, Chicago, 1928.

[10] J. E. Rowe, *Introductory Mathematics,* iii. Prentice-Hall, Inc., New York, 1927.

[11] V. H. Wells, *First Year College Mathematics,* iii. D. Van Nostrand Company, Inc., New York, 1937.

[12] C. S. Slichter, *Elementary Mathematical Analysis,* viii. McGraw-Hill Book Company, Inc., New York, 1918.

The review chapter on elementary algebra has been greatly enlarged. This material is placed in the last chapter or appendix where the considerable amount of very elementary mathematics will not at once confront and perhaps discourage the well-prepared student.[13]

Applications and Illustrations

Variation is made in the content by the differences in application, illustration, and exercises. For example:

Abstract discussions are avoided and frequent applications and illustrations are given.[14]

. . . Illustrations from science are freely used: . . .[15]

While some maturity of the student is assumed, the exposition of the text is in simple language and is complete with illustrations.[16]

An important feature of the book is the large number of graded exercises.[17]

Traditional Content

The content of the preparatory type of general mathematics textbook, however, is about the same as the traditional course, according to many authors.

Although these new features have been introduced, we have adhered closely to the other subjects usually given during the first year of college.[18]

The material in the usual courses in these subjects which is essential for the study of mathematics in the sophomore year is included; and, in addition, there is a rather extensive chapter on the derivative and its applications, a chapter on empirical equations, and one on the foundations of algebra.[19]

[13] Ibid., v.

[14] F. S. Woods and F. H. Bailey, A Course in Mathematics, v. Ginn and Company, Boston, 1907.

[15] C. S. Slichter, Elementary Mathematical Analysis, vii. McGraw-Hill Book Company, Inc., New York, 1914.

[16] V. H. Wells, First Year College Mathematics, iii. D. Van Nostrand Company, Inc., New York, 1937.

[17] Ibid., iv.

[18] J. E. Rowe, Introductory Mathematics, iv. Prentice-Hall, New York, 1927.

[19] Reprinted by permission from A First Year College Mathematics, by H. J. Miles, v, published by John Wiley and Sons, Inc., New York, 1941.

Prerequisites Assumed

The preparatory textbook, in general, indicates that a minimum preparation of elementary algebra and plane geometry is essential for success in the subject.

It presupposes on the part of the student only the usual minimum entrance requirements in elementary algebra and plane geometry.[20]

It therefore presupposes the essentials of high-school algebra, of plane and solid geometry, and of trigonometry.[21]

Time Needed to Meet Objectives

The authors of the preparatory textbooks usually state that the material of the textbook is organized for a five-hour course for both semesters of the freshman year, but suggest that the time may be varied slightly if certain topics are omitted. Typical of such claims are the following:

The material can be covered without systematic omissions in a course which devotes five hours per week for one year to the study of mathematics.[22]

Because time is saved by concentrating on the essentials and by taking advantage of the interrelations of the subject matter, this course can be satisfactorily taught in the time usually devoted to separate courses in algebra, trigonometry and analytic geometry.[23]

The instructional materials constitute a five-hour course for two semesters. Practice has shown that the material of Chapters I–VI can be adequately covered in one semester, and the rest, in a second semester. The abundance of materials makes a modification of the above program possible.[24]

[20] J. W. Young and F. M. Morgan, *Elementary Mathematical Analysis,* v. The Macmillan Company, New York, 1917. Quoted by permission of the publisher.

[21] E. R. Breslich, *Correlated Mathematics for Junior Colleges,* ix. The University of Chicago Press, Chicago, 1928.

[22] L. C. Karpinski, H. Y. Benedict, and J. W. Calhoun, *Unified Mathematics,* iv. D. C. Heath and Company, New York, 1918.

[23] Reprinted from *A First Year College Mathematics,* v, by H. J. Miles, published by John Wiley and Sons, Inc., New York, 1941.

[24] J. S. Georges and J. M. Kinney, *Introductory Mathematical Analysis,* v. The Macmillan Company, New York, 1938. Quoted by permission of the publisher.

In a four-hour course there are certain omissions which can be made by the teacher at his own discretion; the three chapters on solid analytical geometry are not commonly presented in the ordinary four hour course; the chapter on "Poles and Polars" may also be omitted.[25]

There is sufficient material for a year and a half course three times a week, enabling the instructor to begin at a place suitable to his students' preparation and to choose those topics most useful to his class. Certain chapters and sections are marked with an asterisk to indicate that they may be omitted.[26]

Summary of Provisions for Meeting the Objectives as Indicated by the Authors of Preparatory Textbooks

Thus in the field of the preparatory type of general mathematics the authors have indicated an attempt to emphasize an understanding of the concepts as well as the ability to manipulate mathematical symbols. They have endeavored to present large mathematical ideas that disregard compartmental lines. The reorganization has varied the sequence of the topics from that of the traditional order of presentation. However, the authors have indicated that the content includes the same topics as those found in the traditional course it attempts to replace. The authors also suggest prerequisites and time for the completion of the course that are identical to those of a traditional freshman mathematics course.

PROVISIONS AS INDICATED BY THE AUTHORS OF CULTURAL-PREPARATORY TEXTBOOKS

Emphasis on Meaning

In like manner a survey was made of ways in which the authors of the cultural-preparatory type of textbook expected to attain the objectives of their courses. An understanding of the mathematical principles as well as skill in mathematical techniques was found to be emphasized. This emphasis, as will be

---

[25] L. C. Karpinski, H. Y. Benedict, and J. W. Calhoun, *Unified Mathematics,* iv. D. C. Heath and Company, New York, 1918.

[26] C. H. Helliwell, A. Tilley, and H. E. Wahlert, *Fundamentals of College Mathematics,* v. The Macmillan Company, New York, 1935. Quoted by permission of the publisher,

noted in the following quotations, is similar to the emphasis given by the authors of the preparatory type of general mathematics textbook.

Among such principles, which have guided us in the preparation of this text, are the following:

1. More emphasis should be placed on insight in and understanding of fundamental conceptions and modes of thought, less emphasis on algebraic technique and facility of manipulation. The development of proficiency in algebraic manipulation as such we believe has little general educational value. . . . A certain amount of skill in algebraic reduction is, of course, essential to any effective understanding of mathematical processes, and this minimum of skill the student must secure. But it seems undesirable in the first year to emphasize the formal aspects of mathematics beyond what is necessary for the understanding of mathematical thought. This is especially true for that great majority of students who do not continue their study of mathematics beyond their freshman year and who study mathematics for general cultural and disciplinary purposes. It seems to us altogether probable, however, that even in the case of students who expect to use mathematics in their later life work (as scientists, engineers, etc.) greater power will be gained in the same length of time, if their first year in college is devoted primarily to the gaining of insight and appreciation rather than technical facility. Experience has shown only too conclusively that in many cases the emphasis usually placed on formal manipulation effectually prevents the development of any adequate sort of independent power.[27]

If this broad cultural aim is accepted as one of the fundamental principles in the selection of material, we shall readily agree that much that is now generally considered necessary can and should be eliminated from our general courses in mathematics. Almost all of the conventional course in solid geometry would fall in this category, as well as much of what is now taught as college algebra, all of the more specialized portions of analytic geometry, etc. The time thus gained could then be used for topics that are culturally more significant. . . . The disciplinary motive for the study of mathematics is the one most often emphasized and need not be elaborated here. . . . We firmly believe that faith in the disciplinary value of mathematics is fundamentally sound. Teachers of mathematics

[27] J. W. Young and F. M. Morgan, *Elementary Mathematical Analysis*, v, vi. The Macmillan Company, New York, 1917. Quoted by permission of the publisher.

need, however, to formulate with precision their aims and purposes in this direction and make their practice conform to this formulation. The disciplinary value of mathematics is to be sought primarily in the domain of thinking, reasoning, reflection, analysis; not in the field of memory, nor of skill in a highly specialized form of activity. . . . Suffice to remark here that the purpose of technical facility is to economize thought rather than to stimulate it. If our primary purpose is to stimulate thought, we must place the major emphasis on the mathematical formulation of a problem and on the interpretation of the final result rather than on the formal manipulation which forms the necessary intermediate step; on the derivation of a formula rather than merely on its formal application; on the general significance of a concept rather than on its specialized function in a purely mathematical relation.[28]

This coherent treatment has many advantages over the three-unit type of course. One of the most obvious advantages is its greater economy, not only in printing and cost of the book, but in its use for study and reference. . . . But even greater advantages are secured in the way of superior understanding of the essential unity of the material.[29]

One of the main guiding principles has been clearness of presentation. The theory is carefully developed and is then immediately applied to typical examples so that the student may readily see the application of the theory to the exercises which follow.[30]

Disregard of Compartmental Lines

According to the authors of the cultural-preparatory textbooks, the compartmental lines of the subject are broken down in a manner similar to that of the preparatory texts. Many authors agree in substance with the principles expressed below:

In the past few decades there has been a pronounced change in the teaching of mathematics. Especially is this true in the secondary schools, where the old order of treating each subject as an entity in itself has given way to simultaneous treatment and correlation of

[28] J. W. Young and F. M. Morgan, *Elementary Mathematical Analysis,* vi, vii. The Macmillan Company, New York, 1917. Quoted by permission of the publisher.

[29] R. W. Brink, *A First Year of College Mathematics,* v. D. Appleton-Century Company, Inc., New York, 1937.

[30] W. E. Milne and D. R. Davis, *Introductory College Mathematics,* iv. Ginn and Company, Boston, 1935.

the material of related subjects. This movement has but recently extended to our institutions of higher learning. That it will continue to gain favor seems to be a logical conclusion in view of the ultimate aims of our educational system and similar movement in other countries. In the present text the authors have not hesitated to combine subject matter where such combination seemed advantageous, but where advantages were not apparent the traditional order has been followed.[31]

The work is not, however, divided into three units devoted to these three subjects (college algebra, plane trigonometry, and analytic geometry), but is built into a complete unified course.[32]

An earnest attempt has been made to correlate closely the essentials of algebra, trigonometry, analytic geometry, and the elementary calculus.[33]

Unification

The most dominant of the unifying concepts of the cultural-preparatory textbook is that of the function. It is advocated in such terms as the following:

Hence, it has been possible to weave these subjects into a unified whole by the thread of the function idea.[34]

If we desire to enhance the general disciplinary value of mathematics, we will seek out and emphasize especially those conceptions and those modes of thought of our subject which are most general in their application to the problems of our everyday life. It is fortunate for our purpose—and it is probably more than a mere coincidence—that the conceptions and processes of mathematics which most readily suggest themselves in this connection are the same that are suggested by the cultural motive discussed earlier. The concept of functionality and the mathematical processes developed for the study of functions are precisely the things in mathematics that have most effectively contributed to human progress in more modern times; and the thinking stimulated by this concept and these

[31] W. E. Milne and D. R. Davis, *Introductory College Mathematics*, iii. Ginn and Company, Boston, 1935.

[32] R. W. Brink, *A First Year of College Mathematics*, v. D. Appleton-Century Company, Inc., New York, 1937.

[33] W. E. Milne and D. R. Davis, *Introductory College Mathematics*, iii. Ginn and Company, Boston, 1935.

[34] J. W. Lasley and E. T. Browne, *Introductory Mathematics*, v. McGraw-Hill Book Company, Inc., New York, 1933.

processes is fundamentally similar to the thought which we are continually applying to our daily problems. "Functional thinking," to use Klein's famous phrase, is universal. It comes into play when we make the simplest purchase, as well as when we attempt to analyze the most complicated interplay of causes and effects. . . . By making the concept of a function fundamental throughout we believe we have gained a measure of unity impossible when the year is split up among several different subjects.[35]

The authors feel that it is important for the student to recognize the essential unity of the materials and methods of mathematics. To this end not only has all material for a first-year course been collected in one book, but the function concept has been used as a unifying principle and graphic methods have been stressed throughout.[36]

This book surveys and unifies, through the concept function, the materials of the basic collegiate mathematics curriculum—algebra, trigonometry, analytic geometry, and the elements of calculus.[37]

The authors of the cultural-preparatory type of textbook are not in agreement as to the most important unifying theme. An organization from the historical angle is suggested when one author says:

The present text has kept the historical point of view always to the fore.[38]

Others set forth the following suggestions:

The theory and problem material have been selected and organized with reference to a uniform gradation of difficulty, and the order of importance to the student.[39]

[35] J. W. Young and F. M. Morgan, *Elementary Mathematical Analysis,* vii and viii. The Macmillan Company, New York, 1917. Quoted by permission of the publisher.

[36] C. C. Richtmyer and J. W. Foust, *First Year College Mathematics,* v. F. S. Crofts and Company, New York, 1942.

[37] C. W. Munshower and J. F. Wardwell, *Basic College Mathematics,* v. Henry Holt and Company, New York, 1942. Quoted by permission of the publisher.

[38] H. T. Davis, *A Course in General Mathematics,* x. The Principia Press, Bloomington, Indiana, 1935.

[39] C. H. Currier, E. E. Watson, and J. S. Frame, *A Course in General Mathematics,* vii. The Macmillan Company, New York, 1939. Quoted by permission of the publisher.

There is not so much uniformity of opinion, however, in regard to the details of the work of the freshman year, . . . but many wish a single textbook which gives, as this one seeks to give, a definite sequence of topics that shall serve as an introduction to college mathematics. . . . It is being generally recognized that such an all-round view of mathematics can best be introduced through the common use of the science, leaving the rigorous treatment of the abstract theory to be taken up later by those who plan to specialize in the subject. It is upon this principle that the present book is planned; that is, the need for each topic is shown first and the work is then developed by means of applications such as the student is likely to meet in subsequent courses in science, economics, mathematics, or other subjects.[40]

Still others would make the organization of the course around several concepts, as is advocated in the following statements:

The principal unifying ideas are two related concepts, namely, (1) function and (2) the correspondence between geometrical and numerical relations.[41]

Three fundamental concepts governed the selection and organization of the material presented, namely: (*a*) the concept of a function; (*b*) the concept of an equation; (*c*) the concept of a locus. The first has predominated. . . .[42]

In any case most of the authors have advocated some type of unification. In many instances this disposition has caused the introduction of the calculus in the general mathematics course, either as an end in itself or as a part of the correlating concept. The importance of the elements of the calculus is pointed out in such statements as these:

The reference [above] to the general cultural and disciplinary aims of mathematical study at once raises the question as to the selection of the material that is to form the content of the course. The cultural motive for the study of mathematics is found in the fact that mathematics has played and continues to play in increasing measure an important role in human progress. An educated man or woman

[40] G. W. Mullins and D. W. Smith, *Freshman Mathematics*, iii. Ginn and Company, Boston, 1927.
[41] R. W. Brink, *A First Year of College Mathematics*, v. D. Appleton-Century Company, Inc., New York, 1937.
[42] W. E. Milne and D. R. Davis, *Introductory College Mathematics*, iv. Ginn and Company, Boston, 1935.

should have some conception of what mathematics has done and is doing for mankind and some appreciation of its power and beauty. To this end our introductory courses should cover as broad a range of mathematical concepts and processes as possible. In particular, they must not confine themselves to ancient and medieval mathematics. . . . The fundamental conceptions of the calculus must be introduced as early as is feasible in view of the essential role they have played in the progress of civilization.[43]

Early in the book considerable attention is given to the calculus in the belief that many who normally take only one year of mathematics will profit more from the ideas and applications of the calculus than from much of the traditional work usually given. Moreover, the early introduction of the calculus should be a benefit to students specializing in science and engineering. Instead of being treated in one or two chapters and then dropped, the calculus is woven as fully as possible into the fabric of subsequent chapters so that by frequent repetition its concepts become more firmly fixed.[44]

These principles [calculus] are kept before the students during almost the entire year.[45]

Algebraic operations and processes are made meaningful through the early introduction of the simple elements of differentiation and integration.[46]

As in the case of the authors of the preparatory type of general mathematics textbook, the authors of the cultural-preparatory textbooks do not go the entire way in subscribing to the idea of the importance of the calculus as a means of reaching the objectives of the freshman mathematics course. The opinion of one of the authors from the dissenting group is given to illustrate this viewpoint:

Except for the development of the notion of the derivative in connection with the slope of a curve, and a few simple applications to

[43] J. W. Young and F. M. Morgan, *Elementary Mathematical Analysis,* vi. The Macmillan Company, New York, 1917. Quoted by permission of the publisher.

[44] W. E. Milne and D. R. Davis, *Introductory College Mathematics,* iii. Ginn and Company, Boston, 1935.

[45] F. L. Griffin, *Introduction to Mathematical Analysis,* iv. Houghton Mifflin Company, Boston, 1936.

[46] C. C. Richtmyer and J. W. Foust, *First Year College Mathematics,* v. F. S. Crofts and Company, New York, 1942.

problems in maxima and minima, the calculus is not introduced. It is the author's feeling that, for the present at least, the calculus belongs in the sophomore year, and that the important thing in the freshman year is to cover those ideas traditionally treated in algebra, trigonometry, and analytic geometry. However, a preview of the derivative and its interpretation in connection with the slope of a curve greatly helps the student to get a proper start the following year when a more formal treatment is in order.[47]

More Applications—Less Rigor

There seems to be a definite tendency for the authors of the cultural-preparatory type of general mathematics textbook to place less emphasis on mathematical completeness and rigor, and greater emphasis on securing the interest of the student by the use of meaningful applications, illustrations, and exercises. The following quotations represent the opinion of authors with this point of view:

> The chief features of the book are its brevity, the breadth and simplicity of its methods, its selection of subject matter for utility and interest rather than for mathematical completeness, and the careful preparation of its problem material. Elementary matters are reviewed from a new point of view whenever further work is built upon them. . . .[48]

> The chief features of the book are the simplicity of method and the selection and order of presentation of the subject matter. The problem material has been selected with reference to utility and interest rather than mathematical completeness. . . . Practical applications have been kept in mind constantly.[49]

> Besides correlating the material of these subjects the authors have sought to show many applications of mathematics in other fields of endeavor.[50]

> The completeness of treatment has resulted in rather a large volume, but it will be found that in actual practice this large book can

[47] F. E. Johnston, *Introductory College Mathematics*, iii. Farrar and Rinehart, Inc., New York, 1940.
[48] W. R. Ransom, *Freshman Mathematics*, iii. Longmans, Green and Company, New York, 1925. Quoted by permission of the publisher.
[49] C. H. Currier and E. E. Watson, *A Course in General Mathematics*, v. The Macmillan Company, New York, 1929. Quoted by permission of the publisher.
[50] W. E. Milne and D. R. Davis, *Introductory College Mathematics*, iii. Ginn and Company, Boston, 1940.

be taught more quickly and easily and with greater satisfaction to the student than a briefer text which attains brevity by the omission of desirable explanations and illustrative material. . . . The present arrangement arouses greater interest on the part of the student and increases his sense of the utility of the subject matter by emphasizing the applications of results to problems in other portions of the field.[51]

One of our principal objectives has been to write a book that can be read and understood by the student himself, and as an aid in attaining this objective we have worked out numerous illustrative examples throughout the text.[52]

The point of view of the student has been kept constantly in mind. Illustrative examples usually precede the general discussions of new topics.[53]

Independent thinking on the part of the student is encouraged by rather frequent inclusion of exercises which require the students to develop theory or to apply theory in new situations.[54]

Throughout the book word statement problems have been emphasized. . . . An unusual number of problems have been drawn from such fields as business, economics, military science, navigation, psychology and sociology.[55]

Traditional Material

The authors of the cultural-preparatory type of general mathematics course point out that emphasis in certain textbooks has been placed on the following: a unifying principle, special topics deemed important by the author, arrangement of topics that induce a better understanding on the part of the students, or exercises and applications that create an interest. They also suggest that, in spite of the variance in sequence of

[51] R. W. Brink, *A First Year of College Mathematics*, v, vi. D. Appleton-Century Company, Inc., New York, 1937.

[52] E. L. Mackie and V. A. Hoyle, *Elementary College Mathematics*, iii. Ginn and Company, Boston, 1940.

[53] W. W. Elliott and E. R. C. Miles, *College Mathematics—A First Course*, vi. Prentice-Hall, Inc., New York, 1940.

[54] C. C. Richtmyer and J. W. Foust, *First Year College Mathematics*, vi. F. S. Crofts and Company, New York, 1942.

[55] C. W. Munshower and J. F. Wardwell, *Basic College Mathematics*, vi. Henry Holt and Company, New York, 1942. Quoted by permission of the publisher.

topics, the content of the general mathematics course is substantially that of the traditional course which it is expected to replace. The following are typical of the claims of the authors of the cultural-preparatory course on this point:

> Instead of beginning with the conventional formal review of algebra, the plan has been adopted of setting forth clearly the types of work in which the student is likely to need algebra in his subsequent study, and following this with a review that is limited strictly to the essentials of the subject. . . . As the work proceeds, other uses of mathematics naturally present themselves, until the student finds himself in possession of a fair working knowledge of algebra, trigonometry, analytics, and the calculus such as he will need in beginning the study of the various sciences or in the introductory work in courses in distinctively college mathematics.[56]

> The review topics, such as the graphs, simultaneous equations, and the quadratic equation, are interwoven with, and related to, the conic sections and other more advanced material.[57]

> We begin with graphical methods because they afford a simple and interesting means of introducing the function-concept and the big central problems—and also because they tend to develop at the very outset the self-reliant habit of attacking problems by "rough and ready" methods of approximation when no better methods are known.
> Refining the graphical methods leads naturally to the calculus. After some work with this, the need for trigonometric functions is seen, and these are introduced. During the work on trigonometry, analytic geometry, etc., the continuity of the course is preserved by frequent problems which require calculus as well as these other subjects. . . . The order of topics, while unusual—especially in starting calculus before trigonometry—is a natural one.[58]

In the selection of material this book differs from many of its predecessors in the field of unified or general mathematics. It is based on the author's conviction that the best preparation for the calculus is a really thorough grounding in algebra, trigonometry,

[56] G. W. Mullins and D. E. Smith, *Freshman Mathematics,* iv. Ginn and Company, Boston, 1927.

[57] C. H. Currier and E. E. Watson, *A Course in General Mathematics,* v. The Macmillan Company, New York, 1929. Quoted by permission of the publisher.

[58] F. L. Griffin, *Introduction to Mathematical Analysis,* iv. Houghton Mifflin Company, Boston, 1936.

and analytic geometry rather than a superficial study of those subjects together with a smattering of calculus. The means are therefore provided for acquiring quite as great technical skill as can be secured from the old-type courses.[59]

The original plan of presenting algebra, trigonometry, and analytic geometry as a tandem course, to permit adequate preparation in each subject, to permit the use of arithmetic and algebra in trigonometry, and of arithmetic, algebra, and trigonometry in analytical geometry is maintained in the revised edition of this book.[60]

In selecting topics treated in the book, the authors have tried to choose those which are of greatest importance, and to arrange them in a natural sequence. Thus the study of exponents is followed by that of logarithms; tangents to the conics are discussed after the derivative has been defined.[61]

From teaching experience, however, they [the authors] are convinced that a better understanding is gained if these subjects are presented in the traditional order. . . . The arrangement of the material allows certain chapters to be omitted without interrupting the continuity of the text. The instructor thus has greater freedom in choosing topics for a particular course.[62]

1. A review of elementary algebra is included in chapter one. . . .
2. Self-tests are included at the end of each chapter. . . .
3. Algebraic operations and processes are made meaningful through the early introduction of the simple elements of differentiation and integration.
4. Approximate computation has received particular attention . . .

The sections on trigonometry and analytics are arranged so that they may be used for separate courses in these subjects if desired. Chapters XXIV to XXVIII cover miscellaneous topics in algebra from which selections can be made at the option of the instructor. Certain articles which may be omitted without loss of continuity have been marked "optional." [63]

[59] R. W. Brink, *A First Year of College Mathematics,* v. D. Appleton-Century Company, Inc., New York, 1937.

[60] H. L. Slobin and W. E. Wilbur, *Freshman Mathematics,* v. Farrar and Rinehart, Inc., New York, 1938.

[61] M. A. Hill, Jr., and J. B. Linker, *Introduction to College Mathematics,* iii. Henry Holt and Company, New York, 1938. Quoted by permission of the publisher.

[62] W. W. Elliott and E. R. C. Miles, *College Mathematics—A First Course,* v. Prentice-Hall, Inc., New York, 1940.

[63] C. C. Richtmyer and J. W. Foust, *First Year College Mathematics,* v, vi. F. S. Crofts and Company, New York, 1942.

Realizing that many students enter college inadequately prepared, we have devoted the first chapter to a review of elementary algebra and, for the sake of thoroughness, have frequently treated the same topic from different points of view.[64]

Thus, the subjects algebra, trigonometry, and the calculus have received treatment much as in the classical treatment of these subjects. Certain topics from algebra, such as Series, Progression, Permutations and Combinations, Mathematical Induction, etc., which tend to render the study of algebra disconnected, have been omitted.[65]

Prerequisites Assumed

The authors of cultural-preparatory textbooks assume practically the same background preparation on the part of the students as do the authors of the preparatory textbooks. The following quotations are indicative of these prerequisites:

It is assumed that these students have had at least two years of high school mathematics.[66]

The work presupposes only an elementary course in algebra, such as is offered in every high school, a familiarity with the important basal propositions of plane geometry, and a knowledge of the simple rules or formulas of mensuration commonly given in arithmetic or in intuitive geometry.[67]

It presupposes on the part of the student only the entrance requirements in elementary algebra and plane geometry.[68]

The course is adapted to students of widely differing preparations. A knowledge of plane and solid geometry and of algebra through quadratics is the most suitable equipment; but a number of students who had had only two years of secondary mathematics have carried the course very well.[69]

[64] E. L. Mackie and V. A. Hoyle, *Elementary College Mathematics,* iii. Ginn and Company, Boston, 1940.

[65] J. W. Lasley and E. T. Browne, *Introductory Mathematics,* v. McGraw-Hill Book Company, Inc., New York, 1933.

[66] W. E. Milne and D. R. Davis, *Introductory College Mathematics,* iii. Ginn and Company, Boston, 1935.

[67] G. W. Mullins and D. E. Smith, *Freshman Mathematics,* iii. Ginn and Company, Boston, 1927.

[68] C. H. Currier and E. E. Watson, *A Course in General Mathematics,* v. The Macmillan Company, New York, 1929. Quoted by permission of the publisher.

[69] F. L. Griffin, *Introduction to Mathematical Analysis,* vi. Houghton Mifflin Company, Boston, 1936.

While the book assumes only a minimum preparation in high-school mathematics, new materials and the wide selection of problems will offer a challenge to students with the best preparation.[70]

Time Needed to Meet the Objectives

It is quite evident that a minimum of two years of high school mathematics (elementary algebra and plane geometry) is expected by the authors of the cultural-preparatory textbooks. The time required to complete these courses is given as follows:

In a course meeting five times per week throughout the year, there should be ample time. . . . .[71]

The course as given at ———— College takes four hours a week through the year.[72]

The book is designed for a four- or five-hour course given throughout the year. However, it adequately covers the material offered in the usual three-hour year course for freshmen if articles 31, 43, 66, 71, 72, 130 to 132, 136, 137, and chapters VIII, XV, and XVI are omitted.[73]

What the authors hoped to accomplish is to supply enough important and interesting material to occupy any average class, meeting five times a week, for one college year. Doubtless the majority of classes will not cover all the material, thus affording opportunity of selection according to the judgment of the instructor and needs of the class.[74]

While the book is designed for a one-year course meeting five times a week, it is especially adaptable to courses of varying lengths and interests.[75]

[70] C. C. Richtmyer and J. W. Foust, *First Year College Mathematics*, v. F. S. Crofts and Company, New York, 1942.
[71] J. W. Young and F. M. Morgan, *Elementary Mathematical Analysis*, x. The Macmillan Company, New York, 1917. Quoted by permission of the publisher.
[72] F. L. Griffin, *Introduction to Mathematical Analysis*, v. Houghton Mifflin Company, Boston, 1936.
[73] F. E. Johnston, *Introductory College Mathematics*, iii. Farrar and Rinehart, Inc., New York, 1936.
[74] W. E. Milne and D. R. Davis, *Introductory College Mathematics*, iv, v. Ginn and Company, Boston, 1935.
[75] R. W. Brink, *A First Year of College Mathematics*, vi. D. Appleton-Century Company, Inc., New York, 1937.

All the material can be covered in one year three times a week.[76]

Instructors who do not wish to introduce the ideas of calculus in such a course will find ample material . . . for a six-semester hour course.[77]

The book is very flexible and courses of varied lengths may be easily selected. Enough material has been included for a class meeting *five times a week for one year.* For students having an excellent preparation in algebra only portions of the first eight chapters need be used. If in addition the students have had a good course in trigonometry they may begin with Chapter XVI.[78]

The material of the text is sufficient for the needs of a class meeting five times a week for a college year.[79]

Summary of Provisions for Meeting the Objectives as Indicated by Authors of Cultural-Preparatory Textbooks

Although the objectives given by the authors of the cultural-preparatory type of mathematics course are to prepare both the terminal students and those who expect to pursue further work in mathematics, the methods of attaining these objectives and the content of the course are practically the same as the methods and content given by the authors of the preparatory course: (1) meaning is emphasized in addition to skill in mathematical operations; (2) the function concept is one of the outstanding concepts used in unifying the contents of the subject; (3) the calculus is introduced in the freshman course by the majority of authors, with the minority confining the content material to algebra, trigonometry, and analytic geometry; (4) there is the same tendency to break down the compartmental lines; (5) thought-provoking exercises, applications, and illustrations are emphasized. There is no pronounced difference in the sugges-

[76] E. L. Mackie and V. A. Hoyle, *Elementary College Mathematics,* iv. Ginn and Company, Boston, 1940.

[77] W. W. Elliott and E. R. C. Miles, *College Mathematics—A First Course,* vi. Prentice-Hall, Inc., New York, 1940.

[78] C. C. Richtmyer and J. W. Foust, *First Year College Mathematics,* vi. F. S. Crofts and Company, New York, 1942.

[79] C. W. Munshower and J. F. Wardwell, *Basic College Mathematics,* v. Henry Holt and Company, New York, 1942. Quoted by permission of the publisher.

tions of the authors of the dual objective textbooks and those of the authors of the preparatory objective textbooks in regard to either content or organization of material, except for perhaps slightly less emphasis on mathematical "completeness and rigor" by the cultural-preparatory authors.

PROVISIONS AS INDICATED BY THE AUTHORS OF CULTURAL TEXTBOOKS

Emphasis on Meaning

The question naturally arises, "Do the authors of the cultural type of textbook suggest a different content and organization of material?"

A careful study of the writings of the authors of the cultural type of general mathematics textbooks reveals that they, too, emphasize the value of meaning. But technical facility in mathematical computation is a secondary consideration and of value only as it is needed in understanding the mathematical concepts being presented. One author states:

> It has never been doubted that students of the sciences need intensive training in mathematical techniques. But it is more and more generally admitted that students of the arts and social sciences have little need for such technical skill, save for exceptional cases, such as the few who want to use and understand statistical methods.[80]

The primary objective of cultural mathematics is to develop meaningful understanding and appreciation of the mathematical concepts and principles that have influenced the growth of civilization. A mathematics professor who is the head of the department of mathematics at a large college remarked that the following quotation expresses his view on this point:

> The author has intended to present a course in mathematics which will emphasize the distinction between familiarity and understanding, between logical proof and routine manipulation, between a critical attitude of mind and habitual unquestioning belief, be-

[80] M. Richardson, *Fundamentals of Mathematics*, v. The Macmillan Company, New York, 1941. Quoted by permission of the publisher.

tween scientific knowledge and both encyclopedia collections of facts and mere opinion and conjecture, and which will give the student a wholesome appreciation of the nature and importance of mathematics.[81]

Even the group which does not go all the way in subscribing to this principle places less emphasis on techniques and more on ability to understand mathematics in a broad sense. For example, ten years ago one author wrote:

As far as possible an effort has been made to have one section motivate the next, i.e., to present the doctrine as a related whole rather than as a collection of more or less independent topics.[82]

Unification

The policy of the authors of the cultural textbooks has not been to emphasize the organization of the content material around the function concept to the extent advocated by authors of the preparatory or cultural-preparatory. Other methods of unification have been more prominent. Certain authors have advocated the historical approach as the natural and most satisfactory method of unifying the content.

The order in which topics are presented is generally the order in which these subjects were developed historically. This has been done not only to achieve a natural and logical presentation of the ideas of elementary mathematics, but also to give the student an appreciation of the development of mathematics from its original empirical state to its modern abstract developments.[83]

The aim of the book is to present certain topics of mathematics from the historical and logical point of view.[84]

The masters should be allowed to speak for themselves. . . . excerpts from the original creators are quoted. Growth can be followed in the original sources more virile by far than gossipy para-

---

[81] *Ibid.,* vii, viii.

[82] M. I. Logsdon, *Elementary Mathematical Analysis,* Vol. I, v. McGraw-Hill Book Company, Inc., New York, 1932.

[83] H. R. Cooley, D. Gans, M. Kline, and H. E. Wahlert. *Introduction to Mathematics,* iv. Houghton Mifflin Company, Boston, 1937.

[84] J. H. Zant and A. H. Diamond, *Elementary Mathematical Concepts,* Foreword. Burgess Publishing Company, Minneapolis, Minnesota, 1941.

phrases. In this way the reader can see concepts emerging—as fresh today as when they, too, were "modern." [85]

One author of a cultural type of general mathematics textbook has made the historical approach secondary in his method of organization. He states:

> Previous attempts to use the function concept, graphical representation, etc., as the unifying theme have met with varying degrees of success. I believe that the most promising, natural, and genuine unifying theme has been largely overlooked. The statement that mathematics is basic to all sciences is correct partly in the sense of mathematics as the science of space and quantity, but even more, and much more profoundly, in the sense of mathematics as the totality of logical (hypothetico-deductive) systems and their applications. The ideas of logical reasoning and logical structure are common features of all subjects forming part of the search for truth, to a greater or smaller degree depending on the stage of evolutionary development reached by the individual subjects. [86]

While the main emphasis is on reasoning and ideas, the book is constructed on broad historical lines. The author has attempted to include enough historical and biographical remarks to give the student a feeling for the evolutionary growth of the subject in response to human needs, for the fact that its progress is due to the efforts of human beings, and for the fact that it is still a living subject at which living human beings work. However, strict chronology is often sacrificed in the interest of logical presentation. Neither time nor our purpose would permit an attempt to follow all the blundering efforts of the human race to develop a satisfactory mathematics. [87]

In spite of the fact that some authors contend that "If a 'survey' course is to be more than a miscellany, some unifying theme is required," [88] no attempt is made by a few of the authors of the cultural type of textbook to organize the entire content of the course around a single principle or concept. One author of a late textbook in this field says:

[85] D. Harkin, *Fundamental Mathematics*, vii, viii. Prentice-Hall, Inc., New York, 1941.

[86] M. Richardson, *Fundamentals of Mathematics*, viii-ix. The Macmillan Company, New York, 1941. Quoted by permission of the publisher.

[87] *Ibid.*, viii.

[88] *Ibid.*

Although no attempt is made to employ a single unifying theme for the whole book (unless it be the beauty and charm of the subject), number and postulate and form dominate much of the first part, and the function and limit concepts preempt the spotlight in the latter part.[89]

Traditional Material

The degree to which a unifying principle is used does not seem to be the sole criterion for selecting the content material. Two more popular criteria of the authors of the cultural type of general mathematics textbook are (1) those important concepts of mathematics that require little previous mathematical training to be understood and (2) the mathematics that is "useful" to all citizens.

The traditional topics or their "logical" arrangement seems to receive little consideration by the authors as they select the basic material. One author states:

Basic material is selected; it is developed in homogeneous units, either single chapters or sequences of chapters, with much attention to the interconnections so often neglected and so important to true appreciation.[90]

But this "basic material" is not necessary material required for the further study of mathematics, or material of the usual freshman course in mathematics. In the selection of the content material, one author states:

Many considerations have entered into the choice of material for this course. No topic was chosen simply because of some special appeal to the author or because of a traditional prejudice among mathematicians in favor of it. In truth, all material introduced was examined critically for its possible value to the non-specializing student; this is said with knowledge of the indefinite state of the theory of value. Suggestions were obtained from both natural and social scientists, and the views of mathematicians have been freely solicited. Much material proposed for the course was discarded when it was found that the student was unable to appreciate its sig-

[89] Reprinted from *To Discover Mathematics*, viii, by G. M. Merriman, published by John Wiley and Sons, Inc., New York, 1942.

[90] Reprinted from *To Discover Mathematics*, viii, by G. M. Merriman, published by John Wiley and Sons, Inc., New York, 1942.

nificance. Moreover, a serious attempt has been made to present the work in such a manner that student interest will be maintained. The order of topics is the result of much thought and experimentation, and is regarded by the author as pedagogically sound.[91]

Another author says:

. . . topics are selected that have played and still give evidence of playing an important part in the development of science. The principal aim has been to include material which is both useful and interesting.[92]

The tendency to select the topics from the fields of higher mathematics is emphasized by such statements as the following:

The book expresses the belief that it is not necessary to wait until the graduate school is reached before one can learn something about the more interesting and broadly significant parts of mathematics, that many of the ideas which are involved in advanced parts of the subject can be made accessible to those who bring only a small amount of technical knowledge, provided they also contribute a readiness to concentrate and a taste for abstract thinking.[93]

In the compass of so brief a volume there can only be snapshots, not portraits. Yet, it is hoped that even in this kaleidoscope there may be a stimulus to further interest in and greater recognition of the proudest queen of the intellectual world.[94]

Even though the topics are selected from those branches of mathematics not usually studied by college freshmen, an attempt is made to keep to a minimum the number and difficulty of the mathematical skills and techniques needed to understand these concepts. This attitude is reflected in such statements as the following:

The general method of achieving these objectives is to draw on the entire field of mathematics, ancient and modern, and to present,

[91] C. V. Newsom, *An Introduction to Mathematics*, Introduction. The University of New Mexico Press, Albuquerque, 1940.

[92] D. Harkin, *Fundamental Mathematics*, viii. Prentice-Hall, Inc., New York, 1941.

[93] A. Dresden, *Invitation to Mathematics*, viii. Henry Holt and Company, New York, 1936. Quoted by permission of the publisher.

[94] E. Kasner and J. Newman, *Mathematics and the Imagination*, xiv. Simon and Schuster, New York, 1940.

in a unified manner, many of its major ideas and their significance in other fields of knowledge, without discouraging the student with unnecessary techniques.[95]

There is little use in grumbling about the average student's lack of skill in mechanical routines or in rushing him into further ill-understood techniques. Instead, we accept the student, and his preparation as incontrovertible data and begin by discussing the reasonable character of the elementary mathematics which he formerly knew largely by rote. After a brief introduction to the essential logical ideas which are fundamental to any appreciation of mathematics, the evolution of the number system and the essentials of elementary algebra are discussed from a mature and reasonable point of view. After this foundation is laid, the book provides an elementary but critical introduction to several of the most important branches of modern mathematics, without, however, pushing any chapter so far as to strain the student's technical equipment.[96]

In addition to selecting concepts that require little mathematical technique to be understood, certain authors of the cultural-general mathematics textbooks have endeavored to select topics useful as tools to the majority of students. For example, to justify including statistics as a topic to be studied by these terminal students, one author says:

The course begins with a study of some of the elementary principles of statistics. It is practically impossible to read intelligently today without an understanding of graphs and statistics. Our government reports are usually in statistical form. The intelligent citizen should be able to read these understandingly. Our local school reports require statistics—these must be interpreted. Teachers should be able to make statistical reports. We feel justified in including some discussion of statistics. Our national and our local financial structures at present are so important that some understanding of the mathematical theory of finance is necessary for good citizenship. Bonds and sinking funds and annuities are terms that should be familiar to every college graduate. Life insurance concerns a large proportion of our population so that the elements of this theory should be understood by every student.

[95] H. R. Cooley, D. Gans, M. Kline, and H. E. Wahlert, *Introduction to Mathematics,* iv. Houghton Mifflin Company, Boston, 1937.
[96] M. Richardson, *Fundamentals of Mathematics,* vi. The Macmillan Company, New York, 1941. Quoted by permission of the publisher.

An introduction to trigonometry is included because of its use in other fields. This could be omitted by a large group of students, but it is essential to others. Hence, the number of lessons assigned, though small, seems to be justified.

For psychological reasons we have left the review work in algebra for the last, though logically it should come first.

The review includes some of those principles of high school arithmetic and algebra which seem most essential in later work. The treatment is not exactly as it was in the earlier work and even the bright student will find that it will give him ample exercise for his mind.[97]

Other authors state:

Accounting has now become a flourishing profession, reaching into nearly all the nooks and crannies of public and private business. Members of this profession do not go far in practice without running into a multitude of problems that cannot be readily solved by arithmetic. Economists, biologists and scientific workers in all fields have constant need of the simple elementary concepts that are elaborated in the present volume.[98]

Part I is chiefly concerned with some topics in arithmetic, and Part II is essentially an elementary exposition upon functional relationships.[99]

The general plan of each chapter is to present a discussion of the topic at hand, and then give a few of its historical connections, descriptive material about it for orientation purposes, numerous illustrations of how it enters into the work of the world, and a development of methods of solution.[100]

Mathematical Rigor

With such a selection of content we would expect less mathematical rigor or completeness. In fact, many authors definitely state that mathematical rigor or completeness is not the aim of

---

[97] R. P. Stephens, D. F. Barrow, and W. S. Beckwith, *Freshman Mathematics*, Preface. Division of Publications, The University of Georgia, Athens, Georgia, 1936.

[98] Justin H. Moore and Julio A. Mira, *The Gist of Mathematics*, vii. Prentice-Hall, New York, 1942.

[99] C. V. Newsom, *An Introduction to Mathematics*, Introduction. The University of New Mexico Press, Albuquerque, 1940.

[100] F. W. Kokomoor, *Mathematics in Human Affairs*, v. Prentice-Hall, Inc., New York, 1942.

the course, but they do expect the material to be mathematically and logically sound. A few quotations will illustrate this conviction.

The attempt is made to present a critical treatment of mathematical ideas *at the student's level;* complete logical rigor, according to present standards, is to be neither expected nor desired at this level.[101]

An attempt has been made to use the simplest methods . . .[102]

We have tried to make the material in this book historically and logically sound as well as simple enough to be grasped by beginning college students of the type mentioned above.[103]

As the author reviews the experience of the last eight years, it is interesting to observe how different this course is from the one he first conceived. Actual experience has taught that even the good student in his first year of college mathematics is not sufficiently mature in his thinking to appreciate much exactness.[104]

Applications and Exercises

Some of the authors of the cultural type of general mathematics textbooks have placed little emphasis on formal exercises. In fact, with one author,

The exercises have been relegated to the appendix so that continuity of reading may not be interrupted.[105]

On the other hand, there seems to be the feeling among the majority of authors of this type of textbook that some exercises in manipulation are conducive to more effective learning of the larger concepts. The following quotations are indicative of this point of view.

---

[101] M. Richardson, *Fundamentals of Mathematics,* viii. The Macmillan Company, New York, 1941. Quoted by permission of the publisher.

[102] D. Harkin, *Fundamental Mathematics,* viii. Prentice-Hall, Inc., New York, 1941.

[103] J. H. Zant and A. H. Diamond, *Elementary Mathematical Concepts,* Foreword. Burgess Publishing Company, Minneapolis, Minnesota, 1941.

[104] C. V. Newsom, *An Introduction to Mathematics,* Introduction. The University of New Mexico Press, Albuquerque, 1940.

[105] Quoted from *To Discover Mathematics,* viii, ix, by G. M. Merriman, published by John Wiley and Sons, Inc., New York, 1942.

Just as McGuffey's epoch-forming *Eclectic Readers* taught people to read by giving them something worth reading, so mathematics is best learned by working problems worth working. Such problems are inherently interesting. They have a sort of universal property; not only do they lead to their own proper answers; they also reveal more general relations that open up entirely new fields.[106]

The author believes firmly that, with no more background than 2 or possibly 2½ units of high-school mathematics, worth-while results cannot be attained by *talking about* the topics selected for consideration; consequently the development of new ideas is made to proceed slowly and logically and is accompanied at all stages by numerous exercises and varied illustrations.[107]

Then follows a study guide containing suggestions, questions, and problems; and a set of fifty multiple-choice exercises, each having five possible answers, one of which is correct.[108]

It is expected that the student will consider most of the exercises, as they are important in the development of the argument.[109]

The authors have called attention to the attempt to secure exercises and applications within the knowledge of the student rather than just to illustrate a mathematical principle. One author states:

Applications are discussed throughout, but the discussion is restricted to applications which are within the student's grasp rather than to allow it to degenerate into a Sunday Supplement article on the Wonders of Science.[110]

Style of Presentation

With the selection of different content material has come a different treatment of the material. One author says,

[106] D. Harkin, *Fundamental Mathematics*, vii. Prentice-Hall, Inc., New York, 1941.

[107] M. I. Logsdon, *Elementary Mathematical Analysis*, Vol. I, v. McGraw-Hill Book Company, Inc., New York, 1932.

[108] F. W. Kokomoor, *Mathematics in Human Affairs*, v. Prentice-Hall, Inc., New York, 1942.

[109] C. V. Newsom, *An Introduction to Mathematics*, Introduction. The University of New Mexico Press, Albuquerque, 1940.

[110] M. Richardson, *Fundamentals of Mathematics*, vi. The Macmillan Company, New York, 1941. Quoted by permission of the publisher.

. . . a somewhat unconventional treatment has seemed to be in order.[111]

Another author in this group says:

The book, then, cannot be written in formal (and formidable) "textbook" style, and I have chosen to attempt an informal, sometimes chatty, exposition often verging on the narrative.[112]

The authors have pointed out that, although the style of writing might be more interesting and entertaining, the difficult concepts have not been avoided just to select easy material. Even though the style and presentation might make an easier approach to mathematics—as one author says: "To study mathematics is to learn a new language. Just as children learn their mother tongue almost without effort, so it is possible to set forth step by step an easy approach to mathematics." [113] The primary objective is not to create "merely an expository essay, nor an attempt to provide a sugar-coated and worthless course for incompetents." [114]

Concerning this attempt to provide worth-while mathematical material, one author says:

The style of writing, admittedly a radical departure from the usual formal exposition of a textbook, is chosen as an auxiliary to assimilation—not sugar coating, but a digestive stimulant.[115]

Another author warns:

But whatever the surface novelty of treatment, difficulties have not been avoided, and reasonable rigor has been preserved.[116]

[111] R. S. Underwood and F. W. Sparks, *Living Mathematics*, v. McGraw-Hill Book Company, Inc., New York, 1940.
[112] Quoted from *To Discover Mathematics*, vii, by G. M. Merriman, published by John Wiley and Sons, Inc., New York, 1942.
[113] Justin H. Moore and Julio A. Mira, *The Gist of Mathematics*, viii. Prentice-Hall, Inc., New York, 1942.
[114] M. Richardson, *Fundamentals of Mathematics*, vii. The Macmillan Company, New York, 1941. Quoted by permission of the publisher.
[115] Reprinted from *To Discover Mathematics*, viii, by G. M. Merriman, published by John Wiley and Sons, Inc., New York, 1942.
[116] R. S. Underwood and F. W. Sparks, *Living Mathematics*, v. McGraw-Hill Book Company, Inc., New York, 1940.

Time Needed to Meet Objectives

The authors of the cultural type of mathematics textbooks usually have arranged and selected enough material for a one-year course meeting three times a week. There are a minority who have planned a longer course or a shorter one-semester course, but the following quotations will indicate the normal time required to complete the textbook of this type.

In a course which meets three hours a week for one year, it should be possible, on the average, to teach in class about three fourths of the material in the book. Some of the omitted topics may be assigned to students as required reading.[117] [At an Eastern college, a cultural type text] . . . has been used in a 3-hour-per-week course lasting two semesters. There is more material in the book than can be covered in such a course. The first term's work has been based on Chapters I to VIII. . . . The second term's work has been based on various selections of material from the remainder of the book.[118]

Prerequisites Assumed

It might be assumed, in the study of the concepts of higher mathematics and important meaningful applications, that a substantial background in mathematics would be a prerequisite, but the authors have indicated that they do not believe this to be the case. One author states:

The notion that these ideas can be grasped only at the graduate school level is pure myth. Moreover, they can be given to freshmen without introducing many complicated skills and techniques. . . . While it is assumed that the student has had some previous acquaintance with elementary algebra and plane geometry, almost no accurate recollection of the details of these subjects is prerequisite for this book.[119]

Another author says:

[117] H. R. Cooley, D. Gans, M. Kline, and H. E. Wahlert, *Introduction to Mathematics*, v. Houghton Mifflin Company, Boston, 1937.

[118] M. Richardson, *Fundamentals of Mathematics*, x. The Macmillan Company, New York, 1941. Quoted by permission of the publisher.

[119] *Ibid.*, vi, vii.

For a long time it has been the belief of the author that a book both *about* and *of* mathematics could be written informally and nontechnically enough to enable a person of average ability and with almost no previous preparation to master it. This book is the outcome of that belief.[120]

Summary of Provisions for Meeting the Objectives as Indicated by the Authors of Cultural Textbooks

Returning to one of our questions, "How do the authors propose to meet their objectives?" we can briefly answer it as follows:

The authors of the preparatory type of textbook propose content similar to that of the traditional courses in freshman college mathematics; in fact the prerequisites are the traditional high school elementary algebra and geometry as minimum preparation. The material of the course is usually organized around a unifying theme. The attempt at correlation may use the historical or the functional approach. The organization may cause the calculus to appear early in the sequence or be postponed until later in the course. In a few cases the calculus is not considered in the freshman year. The meaning of the operations is emphasized as well as the techniques of manipulating algebraic symbols. The authors suggest that in most cases enough material is provided for a five-hour course throughout two semesters. Emphasis is placed on the variety and number of exercises provided for the student.

The authors of the cultural-preparatory type of general mathematics textbooks suggest content similar to that of the preparatory type. A slightly greater number of these textbooks than of the strictly preparatory include the calculus. While mathematics rigor and completeness are acknowledged as desirable, they are not stressed by the authors as being of primary importance. Other characteristics of these textbooks are: The material is usually organized around some unifying principle, the meaning of the operations as well as the ability to perform them is emphasized, the prerequisites require a minimum of

[120] F. W. Kokomoor, *Mathematics in Human Affairs*, v. Prentice-Hall, Inc., New York, 1942.

one year of algebra and one year of geometry, and the authors suggest that the time to complete the course be approximately five hours per semester for the year. The courses, according to the authors, are in substance the same as those proposed by the authors of textbooks for students who expect to continue work in mathematics or science, with the exception that, in some cases, less rigor and mathematical completeness are stressed. In other words, even these authors propose a weak solution of the same thing. Thus, to accomplish the dual objective—cultural and preparatory—the authors advocate content, organization, concepts, emphasis, prerequisites, and time to complete the course similar to those of a preparatory course.

In the case of the cultural type of textbook, the authors also emphasized the necessity of a thorough understanding of the concepts, but instead of meaning *and* mathematical facility, it is meaning first, with little emphasis on the development of the ability to manipulate mathematical symbols. The organization is built less around the function concept of the calculus and more around the historical, psychological, and logical viewpoint. An attempt is made at an organization which will be interesting and appealing to the student. It may be "sugar coating," but the purpose is to aid "digestion."

The content may consist of certain topics from algebra, trigonometry, geometry, the calculus, and higher branches of mathematics, but the traditional subjects are not used as criteria for the selection of the subject matter. Material deemed essential in the equipment of an educated citizen and of such a nature that it can be taught to individuals with little mathematical competency is the criterion generally stated by the authors. The exercises, when given, are for the purpose of explaining concepts and not for drill to perfect manipulative ability. The style of writing is not that of the traditional textbook but one proposed to foster interest and the desire for greater mathematical insight. The authors suggest a three-hour session for two semesters as the time required to complete this course.

Thus we see that, although the objectives may not be the same for the preparatory and the cultural-preparatory courses, the authors suggest similar content, organization, style of writing, prerequisites, and length of course, while, on the other hand, the authors of the cultural type advocate a different content, organization, style of writing, prerequisites, and length of course.

Provisions as Indicated by an Analysis of Textbooks

To further answer the question, "How do the authors of these textbooks propose to meet their objectives?" it may prove helpful to examine the content of some of the textbooks of each of the groups.

More than fifty textbooks[121] in this field were carefully analyzed. The publication dates ranged from 1907 to 1942. The titles of these books were secured from the Publishers' Index, card indexes of the libraries at Teachers College, Columbia University, the mathematics library of Columbia University, and Questionnaire A.[122]

Although the list does not claim to be exhaustive, it may be considered representative of the textbooks used in this field. The contents of the books were analyzed in regard to the proportion of pages and exercises devoted to the topics discussed. It was found that the contents of all the textbooks would usually fall into approximately sixty categories. For convenience these topics were grouped under several main divisions, which were, for the most part, the general topics suggested by the Joint Commission for General Mathematics Type I and General Mathematics Type II.[123]

With the exception of three textbooks, the traditional material of algebra, trigonometry, and analytic geometry was common to all the preparatory type courses. The difference was in the inclusion of the calculus and special topics, and em-

---

[121] See Bibliography.
[122] See Introduction, page 3.
[123] Report of the Joint Commission of the Mathematical Association of America and the National Council of Teachers of Mathematics, *The Place of Mathematics in Secondary Education*, pp. 159-161. Bureau of Publications, Teachers College, Columbia University, New York, 1940.

phasis on certain topics. The three books whose content did not fall in the preparatory pattern consisted of: (*a*) a second volume of a textbook designed for a one-semester course to follow the first volume, a survey text; (*b*) a book originally used by the author as a terminal course to meet the needs of students specializing in a particular vocation (the author suggested that it took the place of algebra and trigonometry); (*c*) a survey type of a course designed for a class meeting three times a week for one semester.

Thus, in those general mathematics textbooks in which the objectives, according to the authors, were of a preparatory nature, the content was substantially that of the traditional freshman course.

The analysis of the content of the cultural-preparatory textbooks revealed content and style of presentation similar to that of the preparatory texts. This was true for all except two books. One of these was written as a part of a two-volume work: Volume I placed greater emphasis on algebra and trigonometry than did the preparatory textbooks, whereas Volume II placed greater emphasis on analytic geometry and special topics of algebra. However, the combined volumes contained the traditional material usually given in the freshman course. The other preparatory textbook was designed for a three-hour course and contained parts of algebra, trigonometry, and business mathematics. But with these exceptions, the content, organization, and style of writing of the cultural-preparatory were similar to the corresponding properties of the preparatory type of textbook.

The analysis of the content of the cultural type of textbook presented a picture different from that presented by the examination of the content of the other two types. Two of the books contained substantially the same topics as the preparatory but upon a slightly more elementary level. One of these was written in the traditional style and the other in a more narrative form. Two other books of the cultural type, designed for a three-hour one-semester course, contained topics from algebra, trigonometry, and business mathematics, written in the traditional textbook manner but upon an elementary level. These four general mathematics textbooks do not seem to be so popular, however, since less than three per cent of the colleges reporting indicated their use.

In general, it can be said that all other cultural textbooks contained less emphasis on algebra, trigonometry, analytic geometry both in proportion to pages and to problems devoted to the discussion of the topics and in addition contained topics from many of the fields of higher mathematics. Greater than even the difference in the content was that of the style of writing. The narrative, informal discussion, at times bordering on the chatty type, seemed to prevail. The desire to secure the interest of the reader rather than to give a logical presentation of mathematics revealed itself in the organization of the material.

Thus the examination of the content of the general mathematics textbooks revealed that the preparatory and cultural-preparatory contained substantially the same material with similar organizations and style of presentation, but differed from the cultural textbooks in content, organization, and style of writing.

Provisions as Indicated by the Teachers

Teachers of Preparatory Mathematics

The aims of general mathematics proposed by the authors of the textbooks and by the teachers of the courses have been considered. It has been observed that the methods and content which the authors suggest should be utilized in meeting these objectives. Perhaps it would be well at this point to consider the ways in which the teachers are attempting to meet the objectives that they have set for themselves.

A study of a questionnaire[124] returned by a large number of mathematics teachers revealed that 60 per cent of the teachers who use a preparatory text do not attempt to cover the entire material of the text. Of the 40 per cent who are trying to cover

---

[124] See Introduction, page 3.

the entire material, nearly a third indicated that certain topics were covered but with little emphasis. The topics in the preparatory text usually omitted or receiving little emphasis are the following: topics not usually found in the traditional subjects of algebra, trigonometry, analytic geometry, general theory of determinants, circular functions, inverse trigonometry functions, trigonometric equations, theory of equation, progressions, interest, permutations, combinations, and probability. In the textbooks where the calculus is confined to separate chapters, usually those sections containing the integral calculus are omitted and in many cases the differential calculus is omitted or receives little emphasis. The topics omitted are not confined to a particular place in the sequence. For example, nearly 80 per cent of the instructors using one preparatory textbook indicated that a topic on the conic sections in the latter half of the book received special emphasis, whereas topics on the calculus and identities near the beginning were omitted or received little emphasis.

The topics in the preparatory type of general mathematics textbook that received special emphasis are simple trigonometric functions, solution of right triangles, solution of oblique triangles, the fundamental operations in algebra, solution of simultaneous equations, solution of quadratic equations, straight line, and conic sections.

All the instructors of the preparatory textbooks reported that the classes were composed of normal or above-normal students. Similarly, all except one instructor stated that "the course was designed primarily for freshmen." The exceptional case included both freshmen and sophomores.

No syllabi were reported to have been used with the preparatory textbooks. However, the use of supplementary mimeographed material was reported in one case, but upon examination it was found that this material consisted of an outline of the assignments in the text for the course. In general, the text was followed without the addition of supplementary material, and in many cases with the omission of the topics that have been indicated. These preparatory-type courses which were designed for classes meeting five times a week for two semesters are in 50 per cent of the cases being covered, with the omissions indicated, in a class meeting three times a week for two semesters. In only 18 per cent of the questionnaires was it reported that it was a ten-semester-hour course. Thus an examination of the questionnaires indicates that the instructors of the preparatory type of general mathematics are following rather closely the textbook being used. The topics which are omitted to fit the text to a shorter course are usually topics not found in the freshman traditional course, and in many cases the authors suggest they may be omitted if a shorter course is desired. The topics stressed are topics which are predominant in the traditional courses. This situation seems to be true, irrespective of the aims and objectives indicated by the teacher.

Teachers of Cultural-Preparatory Mathematics

In the case of the questionnaires[125] returned by instructors using a cultural-preparatory textbook, three fourths of the instructors indicated that they were not attempting to cover the entire material of the text. Many who were attempting to cover the entire material commented that certain chapters and topics received little emphasis. The topics which received little emphasis or which were omitted are complex numbers, curve fitting, circular functions, trigonometric equations, inequalities, parametric equations, exponential and hyperbolic functions, permutations, combinations, probability, tangents and normals, mathematics of business, solid analytic geometry, and practical mensuration.

The topics given special emphasis were items prominent in the traditional courses of algebra, trigonometry, and analytic geometry. They usually consisted of simple trigonometric functions, solution of triangles, linear equations, quadratic equations, solution of oblique triangles, straight line, conic sections, and the four fundamental operations in algebra. There were a small number of topics on which the instructors seemed to be about equally divided in regard to the emphasis given;

[125] Questionnaire A in Appendix B.

these consisting mainly of the calculus, determinants, progressions, and series. It was expected that the instructors who checked the phrase "terminal course primarily for a cultural education" as the best description of their course would emphasize such chapters as statistics and business mathematics probably to a greater extent than those whose objective was preparatory. But this was not the case. There seemed to be little difference in the emphasis and selection of the content material, whether the objective was cultural or preparatory.

Similarly, the order of the topics in the textbook seemed to have little, if any, bearing on the selection of the material. For example, in one book the discussion of complex numbers occurs before the chapter on the straight line, and all the instructors using this book indicated that they omitted or placed little emphasis on complex numbers but stressed the straight line. In another, the order was reversed, but the emphasis was placed on the same topics. The same was true for other combinations, such as solution of triangles and certain topics in the theory of equations, logarithms and mathematical induction, the conic sections and inequalities. Thus the sequence is not a criterion for judging the relative emphasis that is given a topic.

Ninety-five per cent of the instructors reported that the course was designed for freshmen who were normal or above and that a syllabus was not used. In 60 per cent of the questionnaires the instructors indicated that the course covered six semester hours or less.

The report of the instructors using a cultural-preparatory textbook indicated that they are attempting to cover substantially the same content material, irrespective of objectives; and this material is the traditional topics stressed by the instructors of the preparatory textbooks. Likewise, the instructors of the cultural-preparatory type of text agree with the instructors of the preparatory textbook as to the type of student for whom the course is designed; that is, freshmen who are normal and above.[126] However, the cultural-preparatory group of instructors did indicate that they were giving slightly less time to the

[126] See Questionnaire A in Appendix B.

course. Fifty per cent of the preparatory teachers were attempting the course in six semester hours or less, while 60 per cent of the cultural-preparatory were attempting the course in the same length of time. At least the instructors of both groups were in agreement on all the issues discussed, except possibly the time to give to such a course. In any case the content and emphasis of presentation of the general mathematics course indicated by the instructors using the cultural-preparatory type of textbook are either substantially those indicated by the instructors using the preparatory textbook or more elementary aspects of the same material.

Questionnaire from Teachers of Cultural Textbooks

An examination of the questionnaires received from the instructors using a cultural type of textbook revealed that they, in the same proportion as the instructors of the preparatory textbook, attempted to complete the entire textbook and agreed as to the type of students for whom the course is designed (that is, freshmen who are normal or above). In some respects the questionnaires from these two groups differed materially. The time given to the cultural course is usually a class meeting three times a week for two semesters. Only two schools reported a longer time devoted to general mathematics while 20 per cent of the instructors who answered the questionnaire reported less time.

A difference more prominent than that of the time given to the subject was the difference in content emphasized. The topics stressed were types of reasoning, nature of mathematics, rival views, logical nature of mathematics, significance as a system of thought, function concept, number system, unsolved problems, and the development of mathematics. The chapters and topics indicated as being omitted or as receiving little emphasis were concerned with such topics as limits, theory of relativity, the application of mathematics to other fields, critical study of Euclidean geometry, cardinal numbers, classes, and group concepts. Some of the topics that were neither checked as being stressed nor omitted were elementary algebra topics, logarithms, elementary trigonometric concepts, and non-Euclidean geometry.

Thus, according to the instructors of the cultural type of mathematics, the content emphasized to meet the objectives of the course differs materially from the topics considered in the traditional freshman courses. On the other hand, the material emphasized by the instructors of both the preparatory and the cultural-preparatory general mathematics courses to meet their respective aims is substantially the mathematics material of the traditional freshman courses.

PROVISIONS AS INDICATED BY CLASSROOM OBSERVATIONS

In order to understand more fully the ways in which the instructors of the courses in general mathematics were trying to meet their objectives, the investigator observed more than fifty class recitations conducted by thirty-seven different instructors. These recitations were recorded in substance and carefully studied after the class period. In many cases copies of former examinations were secured from the instructors which threw light on the material covered in the course.

Classes Using a Preparatory Textbook

Without exception the classes using a preparatory type of mathematics discussed traditional topics of the freshman year. For example, in a class of twenty students using a preparatory type of mathematics textbook the subject of the ellipse was discussed. The instructor introduced the lesson by reviewing the meaning of locus. He gave the traditional development of the equation of the ellipse. The students responded well to the few questions asked. However, the lesson was mainly of the lecture method. After using xy for any point p, the instructor changed symbols, which confused one girl. The instructor patiently and perseveringly explained the development until the girl apparently understood the procedure. All the students said that they expected to major in mathematics. The class recitation was similar to those that have been recorded in the traditional analytic geometry class.[127] In this class, as was the case in 60 per cent of the classes visited by the author, the instructor gave the next day's assignment after the bell rang for class dismissal.

In another class of twenty-four students, the instructor, when he came into the room, sent approximately half of the students to the board. He gave them numbers to be written in standard form, assisting those who made errors. After one third of the class period had been spent in this way, he briefly explained how to find the characteristic of the logarithm by use of the standard form. The students remained standing at the board during this short discussion. The instructor then asked the entire class to find the characteristic of some numbers by putting them in the standard form. After fifteen minutes of drill in writing the characteristic of numbers, the instructor showed them how to find the mantissa in the tables (no interpolation was required). This explanation was very short and was given while the students remained at the board. The remainder of the class period consisted of blackboard drill in writing the mantissa of numbers given by the instructor. The students at the seats watched the board but worked only a few of the exercises given. The assignment for homework was to solve four problems from the textbook which consisted of drill in putting numbers in standard form.

Another class using the preparatory type of general mathematics text was studying the same topic but employing a different method. When the instructor came into the room he designated students to put on the board problems that had been assigned the day before. Approximately fifteen minutes of the class period was used in placing the work on the board. Each student was asked to explain his problem. There were occasional questions asked by the students. For example, in finding the cube root by the use of logarithms, a student asked, "Is that

---

[127] Joseph Seidlin, *A Critical Study of the Teaching of Elementary College Mathematics,* pp. 21-25. Bureau of Publications, Teachers College, Columbia University, New York, 1931.

$1/3$ of the quantity or is it the quantity over 3?" The instructor gave a brief explanation of the solution.

In finding the cube root of .1730, several students were confused concerning interpolation. The instructor gave a few remarks on approximate computation while one of the students was going to the board to explain a problem. At this time a student asked, "Are logs accurate enough for building purposes?"

Teacher—If it was a bridge carried to two places, it would fall down, but four places would be enough to build a house.

Student—Say, how about the cube root of a minus number?

Teacher—We'll have some of those tomorrow.

Few questions were asked, however, and the greater part of the period was consumed by students' placing problems on the board and then reciting before the class on traditional freshman mathematics. More than half the classes which were observed used a blackboard recitation of this type. It has been reported by others[128] that this is the most common type of procedure in the traditional freshman mathematics classes in college.

Without exception, the observations of the classes using a preparatory type of mathematics, the conference with the instructors and the students, and a study of the tests available showed an emphasis only on the traditional content material. The frequency of the methods of presentation observed was in the following order: blackboard recitation, mechanical recitation over the assigned work in a textbook, and the lecture method.

Classes Using a Cultural-Preparatory Textbook

Observations were made of classes using a cultural-preparatory textbook. The material discussed was topics from the traditional freshman mathematics courses. The methods of presentation were in substance the methods used in the classes using the preparatory textbook. An exception was observed in those

[128] Joseph Seidlin, *A Critical Study of the Teaching of Elementary College Mathematics*, p. 100. Bureau of Publications, Teachers College, Columbia University, New York, 1931.

college classes in which an attempt has been made to segregate students according to mathematical ability. In the classes of slower students there seemed to be an emphasis on the lecture method. However, the content material of the courses for these students comprised the same topics as those studied by the better students. In fact, in one large college the same examination, consisting of traditional mathematics problems, is given to all sections, with a lower passing score for the slow sections. Conferences with the instructors indicated that the same topics were covered in all sections but on a more elementary level in the slow sections. The difference between the content of the courses for students using a preparatory mathematics textbook and the content for students using a cultural-preparatory textbook, even in the slower sections, was in emphasis on mathematical competency rather than subject matter.

Classes Using a Cultural Textbook

The observations of the classes in which the cultural type of textbook was used revealed two apparent innovations: first, many topics not considered in the preparatory type of general mathematics were included; second, the method of presentation in general was different even in the consideration of such traditional subjects as trigonometry, logarithms, and analytic geometry. In addition, class work included discussions regarding number systems other than ours, logic of our algebra, use of trigonometry in science and industry, functional relationship between such items as scarcity of oil and distance from supply, cost of wheat and distance of transportation, etc. Copies of the examinations indicated that emphasis was being placed on material other than the traditional topics of algebra, trigonometry, and analytic geometry.

The method of presentation was inclined to be more of the informal discussion type. More emphasis seemed to be placed on the development of the concepts and less on formal recitation and the reproduction of mechanical problems on the blackboard.

A class using a cultural type of textbook had spent several

days discussing in an informal way the nature of algebra and geometry. During one of these class periods the following conversation between the instructor and several of his students was recorded.

Student$_1$ [129]—You say that algebra and geometry can be made simple, so that normal people, and I consider myself normal, can understand them. I don't believe that trigonometry could ever be made that way. To me trigonometry seems like a bunch of rules and numbers. It does not make much sense. Perhaps I should not bring that up here.

Teacher—I am glad you brought it up. Do you admit that algebra and geometry can be made simple?

Student$_1$—Yes—even geometry.

Student$_2$—Yes—even geometry. (Many students agreed.)

Teacher—But not trigonometry.

Student$_1$—No, I can't understand it.

Teacher—I am sorry that I cannot prove it now. So you will have to just take my word for it. Trigonometry can be made just as simple as algebra or geometry. We have planned to spend our time on other large mathematical problems, but if you like we can go into trigonometry. If you still want to do it—let me know in a few days.

Student$_1$—Do you teach a trigonometry class?

Teacher—Yes, but it is not interesting. We have so many facts to learn in such a short time that you wouldn't like it. I don't even like it.

Teacher—What are we trying to do in this course?

Student$_1$—To show liberal arts students what mathematics can mean to them.

Student$_2$—To see how simple and logical mathematics really is and that normal ordinary people can understand it.

Student$_3$—To see a reason for some of the tricks that we used to do in high school. (These answers were given with enthusiasm and a great interest in the discussion.)

Teacher—Now that we have ironed out some of the dark mysteries of that sign of addition (makes a plus sign on the board) what should we do next?

Student (looking in his notebook)—Take up subtraction. (The students had made an outline of some of the topics that they expected to study.)

[129] Subscripts merely indicate that a different student is speaking and the same subscript may apply to several different students in the class.

Student$_1$—Let's not spend much time on subtraction, so we can come back to the postulates of geometry.

Student$_2$—Can we all make reports tomorrow?

Teacher—Yes, if you like. (On a previous visit many of the students had contributed short three- to five-minute reports.)

Student—Can you tell me the name of a book that would be simple enough for me to understand?

Teacher—Bring your book list to me after class and I'll help you. (The bell rang. The students left the room. One student, however, erased the blackboard while the instructor was helping another student select a suitable book from the suggested list of readings.)

This little glimpse of a class recitation is suggestive not only of the change of content but of the style of class discussion. The content was mathematics, though not necessarily traditional mathematics. It was attacked as an intellectual challenge and not as a number of mechanical puzzles. The recitation was informal but well conducted. Both the students and the instructor participated actively and enthusiastically in the proceedings. The students' questions seemed to be spontaneous. The instructor's remarks seemed to grow out of the situation at hand. Nevertheless, they guided the students' activities in a definite direction.

There was variation from teacher to teacher of the cultural type of general mathematics, but more than half the classes observed were using this method.

The blackboard was used frequently in this method. But instead of the mechanical recitation of a formal assigned textbook problem, the blackboard was used to explain a problem that the students seemed to be interested in. The explanation would be interrupted by questions from the students and in several cases students proposed solutions to the problem other than the one given by the demonstrator. The demonstration of problems was largely voluntary.

SUMMARY OF PROVISIONS FOR MEETING THE OBJECTIVES OF GENERAL MATHEMATICS

The investigation of the provisions for meeting the various general mathematical objectives as indicated by (*a*) the authors, (*b*) the analysis of the textbooks, (*c*) the questionnaires from the instructors in the subject, and (*d*) personal observations in the classrooms of thirty-seven different instructors substantiates the following conclusion:

The provisions for meeting objectives of both the preparatory and the cultural-preparatory mathematics consist of substantially the material of the traditional courses and, in many cases, the arrangement of the topics in the same order. The method of recitation, the style of presentation, and the selection of topics emphasized is that of traditional textbooks.

However, in the cultural type of general mathematics, there has been a departure from the traditional content. The method of presentation is one of psychological informality instead of logical formality. Emphasis by the instructors is on the understanding of concepts and their applications to life situations rather than the explanation of drill problems. The method of teaching shows a tendency to use an informal discussion, with an attempt to secure the voluntary cooperation of the student. The "ground-covering, textbook-repeating recitation, in which students participate technically and mechanically," and the "blackboard recitation, in which students recite and get a grade," often observed in the preparatory and cultural preparatory classes and so pronounced a decade ago,[130] is less evident in the cultural general mathematics classes.

[130] Joseph Seidlin, *A Critical Study of the Teaching of Elementary Mathematics,* pp. 1-100. Bureau of Publications, Teachers College, Columbia University, New York, 1931.

CHAPTER VI

SUCCESS IN MEETING THE OBJECTIVES OF GENERAL MATHEMATICS

Success as Indicated by Authors of Preparatory Textbooks

THE final question to be considered is, "How successfully have the purposes of general mathematics been accomplished?" To arrive at an answer to this question, the investigator secured information from the authors of the general mathematics textbooks, from the teachers of the subject, from observations of class recitations, and from the students studying this type of mathematics.

Although it is not the purpose of this study to criticize any person who is taking part in the attempt to improve mathematical education, it may have seemed at times in recording the facts concerning the evolution of general mathematics that there has been a tendency to criticize the authors of general mathematics textbooks. It is not desired to leave the impression that the authors have quickly compiled a book entitled "General Mathematics," listed objectives applicable to the largest number of students, and without experimentation or forethought placed the book on the market for the use of the neophyte. If there appear to be errors in judgment on the part of the authors, it should be remembered that they are writing for a group entirely different from the select college group of twenty years ago and that they are attempting to solve a huge problem. As the authors of one of the first textbooks state:

So much has been said in recent years in favor of a unified course in mathematics for freshmen that it seems desirable actually to try it out in practice. For this purpose a textbook is necessary. We do not believe that this textbook will solve the problem; the most we can hope for is that we have secured a first approximation.[1]

These "first approximations" are signs of progress and a careful reading of the literature in the field will reveal that they have been the result of many hours of research and experimentation. The following quotations will give some idea of the success of these courses as viewed by the authors of the general mathematics textbooks.

The present work is a revision and rewriting of a preliminary form which has been in use for three years at the ——— University. During this time the writer has had frequent and valuable assistance from the instructional force of the department of mathematics in the revision and betterment of the text.[2]

All the material has been tried out carefully by the author in the classroom.[3]

The book is the result of ten years' experience in teaching a first course covering the topics of this text. It has been used in preliminary form in a total of sixty-four sections by twenty different instructors and has been revised to embody the criticisms made in the course of actual teaching.[4]

Much of the material included in the book has been tested by class-room instruction. This material passed through several mimeographed editions.[5]

The author herewith acknowledges, with thanks, the valuable aid given by members of the mathematics department—who have used the text in mimeographed form for two years.[6]

This book is in no sense an experiment, having been used, in lithotyped form, for the past five years by a number of instructors

[1] J. W. Young and F. M. Morgan, *Elementary Mathematical Analysis*, x. The Macmillan Company, New York, 1917. Quoted by permission of the publisher.

[2] C. S. Slichter, *Elementary Mathematical Analysis*, viii and ix. McGraw-Hill Book Company, Inc., New York, 1918.

[3] E. R. Breslich, *Correlated Mathematics for Junior Colleges*, x. The University of Chicago Press, Chicago, Illinois, 1919.

[4] C. H. Helliwell, A. Tilley, and H. E. Wahlert, *Fundamentals of College Mathematics*, v. The Macmillan Company, New York, 1935. Quoted by permission of the publisher.

[5] H. T. Davis, *A Course in General Mathematics*, x. The Principia Press, Bloomington, Indiana, 1935.

[6] M. Philip, *Mathematical Analysis*, viii. Longmans, Green and Co., New York, 1936.

with many sections of students. It is written after twenty years of experience in teaching college and university freshmen.[7]

[This book] is the result of experimentation to determine the aims of mathematical instruction for college students, the selection and organization of instructional materials for the attainment of those aims, and the methods and modes of instruction and evaluation of instruction. In manuscript form, the text has been taught at ————— Junior College for the past four years. Modifications, revisions, and reorganization of materials have been the direct result of actual classroom instruction.[8]

[This book] (except for Chapter XX) has been used in temporary form at the ————— University for the past two years with excellent results.[9]

SUCCESS AS INDICATED BY AUTHORS OF CULTURAL-PREPARATORY TEXTBOOKS

The authors who have set forth the dual objectives of their textbooks have also indicated that experimentation has played an important part in the organization and selection of the content material. The following statements are from authors of the cultural-preparatory type of text.

The material presented here has been tried out in the class room and by correspondence courses during the past five years. Methods and problems which have proved unsatisfactory have been eliminated.[10]

The two volumes of [this book] are the outgrowth of an experiment which the author has had in progress at the ————— University for the past five years.[11]

The selection of material and arrangement has been revised in accordance with the suggestions that have come to the authors

[7] V. H. Wells, *First Year College Mathematics*, iii. D. Van Nostrand Company, Inc., New York, 1937.

[8] J. S. Georges and J. M. Kinney, *Introductory Mathematical Analysis*, v. The Macmillan Company, New York, 1938. Quoted by permission of the publisher.

[9] Reprinted from *A First Year College Mathematics*, v, by H. J. Miles, published by John Wiley and Sons, Inc., New York, 1941.

[10] C. H. Currier, E. E. Watson, and J. S. Frame, *A Course in General Mathematics*, vii. The Macmillan Company, New York, 1939. Quoted by permission of the publisher.

[11] M. I. Logsdon, *Elementary Mathematical Analysis*, Vol. I, v. McGraw-Hill Book Company, Inc., New York, 1932.

from the many users of the previous edition of the text and from the official readers for the publishers.[12]

In its various stages of preparation and revision it has been used in mimeographed form for the past four years at the University.[13]

The materials presented here have been thoroughly tried out with the freshman classes in —————— College during the past nine years. Problems and methods which have proved unsatisfactory have been eliminated.[14]

The material of this text has been used in planographed form at —————— University for two years, and we desire to express our deep appreciation to all our colleagues for their helpful suggestions given during this time.[15]

The selection of topics has been based on classroom experience over a number of years.[16]

The authors wish to thank all of their colleagues and students who have assisted in the preparation of this book. They wish to express their thanks and appreciation to —————— for careful and accurate preparation of mimeographed copies of the manuscript during its development and use over a period of four years.[17]

The material has been tested over a period of two years, having been used in mimeographed form by a number of different instructors.[18]

Success as Indicated by Authors of Cultural Textbooks

Although the publication dates of the general mathematics textbooks whose authors advocate the cultural objective as the primary aim are all within the last decade, these books are also

[12] H. L. Slobin and W. E. Wilbur, *Freshman Mathematics*, v. Farrar and Rinehart, Inc., New York, 1938.

[13] W. E. Milne and D. R. Davis, *Introductory College Mathematics*, v. Ginn and Company, Boston, 1935.

[14] F. L. Griffin, *Introduction to Mathematical Analysis*, iii. Houghton Mifflin Company, Boston, 1936.

[15] E. L. Mackie and V. A. Hoyle, *Elementary College Mathematics*, iv. Ginn and Company, Boston, 1940.

[16] W. W. Elliott and E. R. C. Miles, *College Mathematics—A First Course*, v. Prentice-Hall, Inc., New York, 1940.

[17] C. W. Munshower and J. F. Wardwell, *Basic College Mathematics*, vi. Henry Holt and Company, New York, 1942. Quoted by permission of the publisher.

[18] C. C. Richtmyer and J. W. Foust, *First Year College Mathematics*, v. F. S. Crofts and Company, New York, 1942.

the result of experimentation and have proved their worth in the classroom, according to the authors. The following quotations will show the manner in which the authors have tested these textbooks to see whether they would fulfill the cultural objective.

I hope that this testing [at a liberal arts college] will have been a sufficient proving ground and that the resulting book may help in some small way to extend the frontiers of appreciating and understanding mathematics.[19]

This opinion [that the topics will not be too difficult for college freshmen] is based on the experience which the authors and other members of the Mathematics Department at ——— have had in teaching the course to over five hundred students, not specially selected.[20]

. . . some fifteen instructors used this book in preliminary editions with about 1,800 students—entirely unselected—through four semesters.

The experience of this large group has confirmed my belief that the subject matter used here is more intelligible, more useful, more appealing, and more appropriate for the freshman student of arts and social sciences than more traditional freshman topics like determinants, synthetic division, complicated trigonometric identities, etc. The fundamental ideas of mathematics constitute a major contribution to human thought and, as such, belong in any liberal education. They can be grasped by freshmen if they are permitted to emerge gradually from simple concrete situations with the aid of common sense as they often did in the actual history of the subject.[21]

From the writings of the authors it would appear that they are not entirely satisfied with their textbooks, but they do feel that the books will make a contribution toward the achievement of the educational objectives that they have set for general mathematics.

[19] Reprinted from *To Discover Mathematics*, viii, by G. M. Merriman, published by John Wiley and Sons, Inc., New York, 1942.
[20] H. R. Cooley, D. Gans, M. Kline, and H. E. Wahlert, *Introduction to Mathematics*, v, vi. Houghton Mifflin Company, Boston, 1937.
[21] M. Richardson, *Fundamentals of Mathematics*, vii. The Macmillan Company, New York, 1941. Quoted by permission of the publisher.

Success as Indicated by the Questionnaire

Opinions of Instructors of Preparatory Textbooks

In exploring the question, "How successfully have the purposes of general mathematics been accomplished?" the investigator not only considered the statements of those who are furnishing the printed material to be used but also sought the opinion of the instructors who are using this material. In this connection the comments that many instructors have made on the questionnaire have proved very helpful.[22] From the group of instructors who used a preparatory type of general mathematics textbook come such statements as the following:

Previous experience with a general mathematics course was disappointing. It was given as a terminal course largely, but now it is given for the first time as a regular mathematics course.

It should be noted that this course is as follows: First semester, algebra and trigonometry; second semester, analytic geometry and a brief introduction to the calculus, and it is not a survey course. For the mathematics majors it has proved satisfactory.

This course has proven satisfactory for our engineering students.

I feel that a unified course is the most reasonable procedure, and the most convenient way of preparing students for the calculus in one year.

I have found this course very satisfactory for engineers. I have not tried it for other types of students.

Our students cover the usual topics in college algebra, trigonometry, and analytic geometry and they seemed to be prepared for their advanced work. (Engineering students.)

The general mathematics that we give here is not the general mathematics as some use the term. It is an integrated course which is really heavier than the course in algebra, trigonometry, and analytic geometry. . . . For our engineers it has provided a sound preparation.

The comments by the instructors of the preparatory type of textbook were, as a group, favorable to general mathematics as

---

[22] Introduction, p. 3. Also Questionnaire A in Appendix B.

a replacement of algebra, trigonometry, and analytic geometry. There were, however, a small group who expressed such reactions as these:

I believe that the subject matter for both our high school and college should be studied. Too many students come to us without foundation in either arithmetic, algebra, or geometry. I do not believe the lack of talent or application on the part of the student is the cause. Nevertheless, many students seem to be hopelessly lost in the course.

In general, I believe the number of persons taking mathematics is on the decline. This is a girls' college, and my class is smaller than usual in general mathematics but slightly larger in the other sections taking the old-time course of college algebra and trigonometry.

I am sorry that the above report is not complete. We have used the present text for the first time this year. We put in the general mathematics course four years ago but did not care for the text which we used then.

Have not found any textbook satisfactory for this course yet. I believe that there is definite need for a general mathematics course organized in somewhat different manner than is now the case.

The text we are using is not very well adapted to our purpose. There is not enough practical work and what there is is not evenly enough distributed.

These quotations from dissatisfied users of a preparatory type of general mathematics textbook represent approximately 16 per cent of the questionnaires.

A further consideration of the questionnaires returned by the instructors using a preparatory type of general mathematics textbook reveals that less than 5 per cent of those whose purposes for the course were preparatory indicated dissatisfaction with the course, and nearly half of this group were using textbooks which the authors suggested were preparatory but the content of which was lacking either in certain topics or in emphasis upon others. One third of the instructors who were using a preparatory textbook but who checked "terminal course primarily for a cultural education" as the phrase that described the course added the information that they were dissatisfied with the course. One fourth of those instructors who indicated that they were using the textbook to meet both a preparatory objective and a cultural objective added the comment that they were dissatisfied with the course.

Thus from the results of the questionnaire it seems that the instructors using a textbook which not only the author has indicated as preparatory but the contents of which are of that type, are in general satisfied with the results of the course. On the other hand, the instructors who are using the preparatory type of general mathematics textbook for purposes other than those suggested by the author are inclined to express dissatisfaction with the course.

Opinions of Instructors of Cultural-Preparatory Textbooks

A critical study of the questionnaires returned by the instructors who use a cultural-preparatory type of textbook suggests that these teachers are not all in harmony regarding the success with which general mathematics is fulfilling the needs of the students. There were some, however, who enthusiastically expressed themselves in favor of the course. This sentiment is indicated by such comments on the questionnaire as these:

I find this very well written . . . meets my problem most satisfactorily. The simplicity, directness, and soundness which the authors have introduced each new topic and the variety of well selected problems fascinate as well as create much interest in my classes.

We are well pleased with the results of the general mathematics course. We also divide the general mathematics students on the basis of ability and previous training; thus we are able to adjust the teaching to various abilities.

The text is satisfactory for the science students in fact, I don't know of its equal in giving a mathematical foundation to science freshmen but for the arts students it is not so satisfactory.

I consider the coordinated course in mathematics for freshmen superior to the separate treatment of college algebra, trigonometry, and analytic geometry usually given in the freshman year.

A course in general mathematics undoubtedly has a place in the very strong liberal arts college, particularly as a terminal course primarily for cultural education.

It is my opinion that the general course has more value for the average student than the conventional course.

We find that our students are much better prepared in every way since we began the unified course instead of the old fashioned airtight compartment methods.

In my opinion, the freshmen who have studied general mathematics became better physics students than those who studied trigonometry and analytic geometry as freshmen. . . . I have taught freshman mathematics many years and general mathematics for the past eight years.

Good course for those desiring only one year of college mathematics. Often students who are not interested in mathematics take this course, and get interested and take more mathematics.

For the student who does not need or desire an extended course in mathematics, I would say the procedure is valuable.

The text is very good . . . gives a glimpse of what more advanced mathematics is about . . . these students do not continue in mathematics but seem to enjoy the course.

The purpose of our course is essentially cultural . . . My reward is that at the end of the year about three quarters of each class think that they have acquired an entirely new conception of mathematics.

I feel that general mathematics is far better than the individual subjects . . . it furnished an opportunity for drill constantly in all phases of mathematics.

Not all the users of the cultural-preparatory type of text expressed themselves as believing that the general mathematics course is meeting their purpose. A number gave lengthy comments either on the questionnaires or in accompanying letters, expressing their dissatisfaction and, in many cases, disapproval of the general mathematics course. A professor who is the head of the mathematics department of a large college wrote, "Count me and my faculty as strongly disapproving of such courses."

The following quotations will serve to indicate the general spirit expressed by the group of dissatisfied instructors.

We are not satisfied with the above text which we have used two years, and expect to make a new adoption altho the text to be used has not been definitely chosen.

The course is open to all freshmen. . . . The text does not fit the course too well and I am looking for a more suitable text which I have not as yet found.

Need more brief historical material *interspersed* in purely mathematical material.

We do not care for the above book. . . . We expect to have two more sections next year than we have now but expect to use a different text.

I speak in first person because I have taught all general mathematics except overflow sections. . . . If I had to judge from available textbooks I would conclude that general mathematics was still in a very unsatisfactory state of development.

Hash—a jumble of so-called practical problems in which you cannot see the forest on account of the trees.

Some authors think you can teach mathematics by *talking about it, stories, wisecracks. Make it interesting a la vaudeville.* It can't be done.

It is my opinion that in such courses students learn too little about too much.

I think this book is good for cultural material and those who do not go on in mathematics. However, it is unsatisfactory for students who wish to go further.

Your questions are a little hard to answer as I change texts each year [A letter accompanies the questionnaire in which the texts used are listed with chapters omitted and stressed]. . . . I still believe in general mathematics but it seems that a text that fits one group is unsatisfactory with other groups.

The course must be organized to take care of students whose preparation is poorer than the preparation of students who took freshman mathematics ten years ago.

We have become convinced that the types of texts used in the so-called unified courses are unsatisfactory. Next year we plan to divide the course into three parts—a two-hour course in algebra to run through the entire year plus a one-semester-three-hour course in trigonometry and a one-semester-three-hour course in analytic geometry.

For the most part we have divided the freshmen into three groups according to their purposes in studying mathematics. . . . [Four page letter describes the groups and reaction to general mathematics.] For this group [those students who do not continue in mathematics beyond the freshman year] I have used a different text nearly every year and have not found one wholly satisfactory. I am looking for a new text for next year.

How to satisfy the needs of each member of the group is an ever present problem.

We used the text with 120 students this year and found it to be deficient. Supplementary material was essential for all scientific and mathematics students. We are going to change texts.

Next year we are going to separate the students into two sections. We plan to use this text for the terminal course but have not decided about the group that goes on in mathematics.

With reference to question 9 (in reference to objectives of course)[23] I should like to say that, as ——— is a small college, it is necessary that this course attempt to be a terminal course for some students, a preparatory course for others. It is not working out well. . . . I plan to change texts next year.

I know of no satisfactory text. But I am convinced of the value of such a course.

The text presupposes a better prepared student from high school than we really get. I wonder if this is the common experience of others.

My main problem is how to determine what material to include when there are *both cultural and preparatory students in the one course.*

Views similar to those just quoted appeared on approximately one fourth of the questionnaires which were received from in-

[23] Questionnaire B in Appendix D.

structors using a cultural-preparatory mathematics textbook. A further study of these questionnaires revealed that 85 per cent of these dissatisfied users were attempting objectives other than, or in addition to, those preparatory for the further study of mathematics. Of the 15 per cent of dissatisfied instructors whose primary objective was preparatory, one third were using a cultural-preparatory type of textbook, the content of which was of the survey type. In other words, only 10 per cent of the dissatisfied comments came from instructors whose objectives were preparatory and who were using a cultural-preparatory textbook composed of strictly preparatory content material. Thus, in the opinion of the instructors, the traditional mathematical material, in general, is meeting the needs of the preparatory students.

A further consideration of this group of dissatisfied users of the cultural-preparatory text showed that one fourth of these instructors had checked "Terminal course for cultural education" as the phrase that best described the course and at the same time were using a textbook whose content was of a preparatory nature. In fact, 90 per cent of the instructors whose aims were cultural but who were using a textbook whose content was preparatory express dissatisfaction with the course.

In general, the questionnaires from the instructors whose purposes were preparatory were favorable toward the cultural-preparatory textbook. However, if the objectives of the instructors were cultural or a combination of cultural and preparatory, then there seemed to be at least a question as to the value of the course.

Opinions of Instructors of Cultural Textbooks

A careful study of questionnaires received from the instructors using the cultural textbooks also showed opposite opinions in regard to the success of the course.

The praise for the cultural textbook was very high on some questionnaires. The following statements taken directly from the replies of the instructors are indicative of the success that some see for the course.

Considerable interest has been aroused by our courses in general mathematics among the students. Registration upon the freshman level has increased nearly 50% in the last five years, and we are convinced that such courses have their place in the educational set-up.

I feel that the course fits the needs of the non-majors admirably. I would even like to see it used for purpose 9c above [preparatory course for further study of mathematics], provided additional hours were added for the development of technical proficiencies.

The text is not entirely satisfactory but I am now convinced of the value of such a course.

Students need such a course. The college should offer a course to such students to show them the significance of mathematics in our culture. Naturally it can't be rigorous but non-rigorous discussions of transfinite cardinal numbers, non-Euclidean geometry, statistics, the function concept, etc., can be made interesting and significant.

The cultural course offered above has proven interesting to students and some have decided to go on in mathematics as a result of awakened interest. A course in cultural mathematics serves to enlighten the student as to the value of and place in society for mathematics. It should be taken by all terminal students.

It is my opinion that this course is far more valuable to many students than the conventional freshman course.

Another questionnaire from a college enrolling more than two hundred and fifty students each year reports,

We have been very much pleased with this course for the purpose for which it is designed and used. I do not feel that it is the best or satisfactory preparation for students who will continue in mathematics or who wish to use it for a tool course.

We have found this course very satisfactory for students whose major interest is in the social sciences and the humanities.

Not all of the instructors were as optimistic about the success of the cultural type of general mathematics. In fact, one third of the users of the cultural type of textbook expressed dissatisfaction with the course. On the questionnaires were such comments as these:

I am rather disgusted with the average textbook which merely condenses the work found in standard books on algebra, trigonometry, analytics, and the like.

We will no longer offer a course in general mathematics. We offered it three years to more than three hundred students. We were not satisfied with the course.

Our institution is organized to operate on the quarter basis and I find the administration of such a course unsatisfactory.

I think I am moving toward the conviction that more exercises are needed to fix habits than the general course offers.

One report from a large college which is now using a cultural type of textbook in attempting to meet all three of these objectives says,

Have used several different texts for about the last ten years. . . . I found that I had to supplement this text a *very great deal*. It was an unsatisfactory text for my students. I am giving up this type of program beginning with next year.

For five years one college enrolling more than two hundred students in general mathematics each year used a cultural textbook to meet a combination of cultural and preparatory objectives. After expressing dissatisfaction with the result of the course, the instructor says:

For this coming year we plan to use this text [cultural text] for group (a) [terminal course for cultural education]. For the other students we are going to resort to courses in algebra and trigonometry in the freshman year and then continue with analytic geometry and the calculus. . . . We are continuing to experiment.

Another instructor in commenting upon the value of the course wrote:

We are fairly well satisfied with the course for those who desire essentially a one-year terminal course. We are not certain that the general course is a substitute for the traditional program in the training of technical students.

The comment of another instructor was:

I am not at all sure that the content of the above textbook is best suited to the needs of students planning to be elementary school teachers.

Approximately one fourth of the comments of the instructors of the cultural general mathematics textbooks were unfavorable to general mathematics as a means of obtaining the aims and purposes they proposed for their classes. A further study of the questionnaires revealed that the instructors who were attempting to meet objectives other than, or in addition to, the cultural objectives were more predominant in the group of dissatisfied users of the cultural type of general mathematics textbook. In fact, of the instructors who reported that their general mathematics course was designed as a "Terminal course primarily for cultural education," less than ten per cent reported that they were dissatisfied with the success of the course, and of this small proportion of dissatisfied users exactly half were using textbooks whose authors indicated that the books were designed for the terminal cultural student but the contents were a weak emulsion of preparatory material, topics common to specialized vocations, or a combination of both.

Thus the questionnaire indicated that those instructors who are using a cultural type of general mathematics textbook for a cultural objective are satisfied that progress is being made. The small number of this group who reported dissatisfaction perhaps indicates a healthy state in the progress of mathematical education in that it shows that this group is experimenting and trying to seek a still better solution to their problems. As one instructor who has more than two hundred freshmen in general mathematics each year wrote on the questionnaire, "I am not entirely satisfied with the course, but it is much better than the old setup [algebra, trigonometry, analytic geometry]. I shall continue to experiment with at least one section."

Summary of Success as Indicated by the Questionnaire

From a study of the questionnaires it appears that, in general, the dissatisfied instructors are not those who are using a preparatory textbook or a cultural-preparatory textbook for preparatory students, nor those who are using the cultural textbook for the cultural terminal student, but instructors who are attempting to meet a combination of objectives or a different objective from that for which the textbook was planned.

Success as Indicated by Class Visitations

Success of Instructors of Preparatory Textbooks

The conference with, and class observations of, those instructors who were using a preparatory type of general mathematics textbook also indicated that in general they are satisfied with the course. Frequently the instructors made such comments as: "It is a much better preparation for the calculus." "The science instructors feel that the students are better equipped." "The students can apply their mathematics much better." "General mathematics furnishes a continuous review of algebra."

There was a difference of opinion in regard to the order of topics. For example, one instructor favored the introduction of the slope of a line before the study of the trigonometric functions, and another instructor in the same school system strongly disapproved. The opinion was about equally divided on the question of whether the calculus should be introduced early in the freshman course or postponed until the sophomore year. The majority of the instructors of the preparatory type of general mathematics textbooks reported in the interviews that the course was more nearly meeting their purposes than the separate courses of algebra, trigonometry, and analytic geometry.

Success of Instructors of Cultural-Preparatory Textbooks

Interviews with instructors who were using the cultural-preparatory texts revealed a variation in the emphasis given to the dual objective of the course. Some instructors frankly stated that the primary purpose of the course was to give a solid foundation in mathematics to those who expected to use it. This group of instructors usually approved the course.

There were, however, instructors who said that an attempt was being made to provide both a cultural and a preparatory education. These instructors, who expected to accomplish a dual objective with the general mathematics course, usually agreed that the study of traditional mathematics was about as good cultural training in mathematics as could be provided. However, when asked if they felt that the course was successful, the answer was usually negative. Typical replies follow: "What can you expect to teach the dull normal?" "The smart ones seem to get it." "The 'dumb' ones [meaning 'dumb' in mathematics] should be in an English or history class. They should not be required to take mathematics." "I just sweat blood trying to get this stuff across, but I feel that I am wasting my time on about half the class." "Half of the class don't expect to use mathematics and are not interested." "If those who were not interested were in a class by themselves maybe we could get some place with them."

This last statement suggests the procedure which at least three colleges are following. The students are enrolled in sections according to their mathematical ability; a cultural-preparatory textbook is used and the same material is covered in all groups except the slower group, with whom the considerations are more elementary. Interviews with instructors indicated that they believe that it was more desirable to segregate the less efficient students but the system was not altogether satisfactory. A few of the remarks typical of the instructors' reactions are as follows:

Since the slow ones must take the course, it gets them all in one section, then they don't all fail. [The passing score was lower for the slow sections.]

It permits us to teach the better students without bothering with the dumbbells.

It puts the headaches all in one or two classes. That's better than having them all day.

It is better than the old set-up, but I don't think the students in the slower sections get much out of it.

I feel that the slow sections get something out of the course but we have not solved the problem.

I may be a "heretic" but I think the only reason the dumb kids don't cry their heads off is because they are not so scared of flunking as when they were not divided according to ability. I think something should be done for them but I don't know what.

Well, after all, I think it is some improvement.

Some of the problems confronting the teachers using a dual objective textbook were very prominent in the observations of the class recitations. After one instructor had given a very enthusiastic and clear but brief explanation of the projection of lines, he gave the students this problem: "A ship is going 15 miles an hour 70° east of north. How far will the ship be in an hour?" About sixteen of the twenty-six students made drawings on coordinate paper which showed the projection of the lines upon the x and y axis. The other students did not understand the meaning of projection. When the instructor observed the difficulty that the slower students encounter he gave them the following instructions: "Don't pay any attention to the projection. Here is a rule you can follow that will give you the same answer. Make a right triangle, using the number of miles an hour as one leg and the number of degrees as the size of the acute angle. Then all you need to do is find the other leg or hypotenuse." Later in the day, in discussing the success of the class, the instructor remarked, "I wish I could have given a better explanation of projections but if I waited for the slow ones we never would cover the material. The better students understand it and maybe the others will get part of the problems right on the examinations."

In a visit to a class using a cultural-preparatory type of textbook, one week before the final examination the following observation was recorded.

The class consisted of sixteen girls and seven boys. The students were very busy working problems when the teacher arrived. The teacher asked certain students to put on the board problems that had been assigned the day before. The problems were formal differentiation and integration exercises.

Teacher—Problem number four is a type that you can expect on the examination. (The students mark it in their books.) Whoever has it, please explain it.

Student—I have that one but I can't get it.

Teacher—Read it for me and I'll help you.

(Student reads the problem $A = \int_1^8 \sqrt[3]{8}\, dx$ and the instructor writes it on the board.)

Teacher—How do we change this problem to integrate it?

Student—Don't know.

Teacher—Some one help us out.

Student$_1$—Change the root to an exponent.

Teacher—You come to the board and show us how to solve this.

(The students integrate the example and substitute the values in the solution.)

Student$_1$—I don't see where he got the four thirds.

Student$_2$—From the rule for integrating.

Student$_3$—I don't see why you put the one third up there instead of the cube root.

Teacher—Just remember you always write it as a power—you had better look this up because you'll have one like it on the examination.

Teacher—Another problem you will have on the examination is like number six. A student has the solution to number six on the board.

(The problem was $A = \int_1^9 \dfrac{1}{\sqrt{x}}\, dx$.)

Student—Where does he get the $(-\tfrac{1}{2})$?

Teacher—The exponent is minus when you take it above the line. Are there other questions?

Student$_1$—I don't see it.

Teacher—If the girl who doesn't see it will come up after class, we will go over it again. (Several exercises in differentiation that were on the board were then discussed by the students. In differentiating the equation $y = e^{-2x^2} + 16$ several questions were asked.)

Student—Where did the $-4x$ come from?

Teacher—By the use of the power formula "$-dv$ is $-4x$."

Student$_1$—Well, what happens to the 16?

Teacher—It dropped out.

Student$_2$—Why do we let the 16 drop out? (The teacher gave a brief explanation.)

Student$_2$—Well, why didn't the 2 drop out?

Teacher—Oh! that's different. Look here. If there is a number such as a 2, 4 by itself with a plus or minus sign before it, then it drops out. If the number is with a letter and no plus or minus sign between them, it doesn't.

Student₂—Oh! I see.

Teacher—Don't forget that. We may not have a problem like that on the examination but we will have one like number 9. Let's get to that one. (Two more problems are mechanically explained by the students without questions from the class.)

Teacher—All right, now number 9. (The teacher observes a mistake in the work.) My! What have you done?

Student—I differentiated.

Teacher—Maybe you *thought* you did but you didn't.

Student₂—She thinks she differentiates e^{3x} like she would $3x$.

Teacher—That seems to be the trouble. How many remember how to differentiate e^{3x}? (About half the students raise their hands.)

Teacher—That's too bad. It's the last one we learned, too. Now listen and I'll tell you how you can keep from becoming confused. The first one we learned was $3x$. Now we will compare that to an automobile because that is our old method and the automobile is the old method of travel. We shall compare e^{3x} with an airplane because that is the last method we learned and the airplane is the last method of travel. Now if we think of x being in the automobile and x being in the airplane then we can see that when x is down on the ground we use the old method of differentiation, but when x is up in the air as an exponent—that is, in an airplane—we use the new method of finding the derivative. (The students laugh.)

Teacher—We only have a few minutes left and I want to show you how to work the problems for tomorrow. Here is a problem. (He wrote the product of two variables.)

Teacher—Now listen and I'll show you a new trick. This is the rule. You write down the first one like this. (He demonstrates on the blackboard.) Then right beside it you write the derivative of the second one. Now don't stop there. Put a plus sign, write down the second variable, and next to it write the derivative of the first variable. (The students seemed to be amazed and several started to ask questions.)

Teacher—Now watch again—sometime later I'll come back and prove this. (Aside to a student, he adds, "If we have time.")

(The instructor repeats the same procedure.)

Teacher—Now let me show you a good way to remember the rule. Think of the two variables u and v as two neighbors. First you write down one neighbor and the differentiate of the other one. But you don't dare to stop there. If you do something to one neighbor and not to another, one neighbor will become jealous; so since you differentiated the second neighbor and wrote down the first, you must now differentiate the first neighbor and write down the second neighbor so they won't be jealous.

Student—Does it make any difference which neighbor is differentiated first? Will that cause a jealous argument?

Teacher—No, it makes no difference. (Bell rings and the instructor makes an assignment of eight formal exercises in differentiating the product of two variables.)

The recorded observation is indicative of the attempt that instructors are making to give, on the one hand, a preparatory course in mathematics to a group who expect to use it as a tool subject, and, on the other hand, some kind of mathematical contribution to the terminal student. After the class left, the instructor frankly stated that he believed that he was not doing justice to either group. He commented that if he spent time with the slow students he didn't have time to give a reasonable explanation for the rules to the good students. He further stated: "I don't like to give those crutches that I did today, but when those poor kids come to class and try; I just can't stand to see them fail. And, after all, I cannot pass them unless they can solve some of the problems on the examination."

The opinion of the instructors using a cultural-preparatory type of textbook was that it is very difficult, if not impossible, to teach a cultural and preparatory course simultaneously if by *cultural* is meant anything in addition to traditional mathematics. The expression, "It can't be done," was common among these instructors.

Success of Teachers of Cultural Textbooks

The picture presented by the visits to the classrooms of the teachers using a cultural textbook was brighter but not one to create extreme enthusiasm.

In general, the instructors expressed themselves as being in favor of a general mathematics course to meet the objectives of a cultural education. The fact that many frankly stated that much improvement could be made is perhaps a sign of healthy growth in this field.

In the class observations, it was noticed that particular emphasis was placed on the understanding of the problems, irrespective of the type of problem being considered. The students contributed problems, some real, some highly artificial. Some of the problems were solved; others were discussed but left unsolved because of lack of technical ability. In general, the work that was done indicated an emphasis on understanding rather than speed in mechanical solutions.

The recorded observation of a class visited the day before the one just reported illustrates this emphasis.

There were twenty-seven students in the class. The teacher walked into the classroom.

Teacher—Will somebody start something? (The instructor adjusts a window to keep the sun from shining on a student and then sits in a chair at the side of the room.)

Student— (without rising) I have a problem here on calculus I would like to have solved.

$Student_1$—I can't hear. Will you stand up?

Student— (Standing, she reads the problem.)

Teacher—How many would like that problem? (Two or three raised their hands.)

Teacher—Let's have another one.

$Student_2$—I have one. (The girl proposed a problem that would require integral calculus to obtain the solution.)

Teacher—That is an interesting problem. We don't have the tools to solve it now. Save it until next week. We may not be able to solve it then but we can discuss it better at that time.

$Student_3$—I have one that I got out of a calculus text. (She gave the familiar telephone problem requesting the number of telephone subscribers necessary for a maximum profit. Several students said that they would like to see a solution to the problem.)

Teacher—Will someone write the problem in terms of symbols on the board? (After considerable discussion by the class the equation $i = (1000 + 100m) (500 - 10m)$ was written on the blackboard.)

Teacher—What is going to change in the equation?

Student—m.

Teacher—Is that all?

Student—Yes.

$Student_1$—No, the i will change.

Student—Sure, I forgot about that side of the equation.

Teacher—What are we trying to find here anyway?

Student—When i will be the largest?

Teacher—How do we do that?

Student—Set the derivative of the right hand side of the equation equal to zero.

Teacher—Then what?

Student$_1$—Solve the equation.

Teacher—Can we take the derivative of the equation, or is it beyond us? How shall we go about it?

Student—We can multiply it out.

Teacher—Will you do it? (Student goes to the board and writes $- 1000m^2 + 40000m + 500000$.

Teacher—Now let's see. Write down the derivative of 500000.

Student—Can't do that 'cause you don't have anything.

Teacher—Why?

Student—500000 is not going to have a rate of change.

Teacher—What will the derivative of 500000 be, then?

Student—Nothing.

Teacher—Perhaps it would be better to say zero rate of change. At least, that is what a mathematician would say. (The student wrote down "0.")

Teacher—What would the rate of change of 40000m be?

Student—40000 times whatever the rate of change is.

Teacher—Can you find the rate of change for the $1000m^2$? (Student turns to the board, draws a little square, and apparently is studying the figure. As the student is writing some numbers on the board in trying to find the solution, a girl raises her hand.)

Teacher—Don't rush him. Let him take his time. If you think you know the answer, check your result. You may be wrong, you know.

Student—It would be 2000 times the rate of change of m times m.

Teacher—How did you get all that?

Student—From the square problem we worked the other day.

Teacher—Will you write down the equation now just like a mathematician? (The student had difficulty in writing the equation.)

$$2000m - 40000 = \frac{di}{dm}$$

(Several students assisted him in arriving at the final form.)

Teacher—If we set the equation equal to zero, who remembers enough high school algebra to solve the equation for m?

Student—I think I can. (He goes to the board and writes $2000m = 40000$.)

Teacher—Why 40000? We have a minus 40000 here.

Student—I changed it to the other side.

Teacher—I see. Can we always do that?

Student—We really don't do that. We added 40000 to each side. (Then the student wrote the addition on the blackboard.)

Student₃—My brother is taking calculus and he didn't know why he changed signs when he took numbers to the other side of the equation.

Teacher—Sometimes we forget. What is the value of m?

Student₁—20.

Teacher—What if m was 19½? (The student looked at the board and seemed puzzled.)

Student₁—Just leave it 19½. (Several hands went up.)

Teacher—What is all this fuss from the class?

Student₂—There couldn't be a half subscriber so I would say 20.

Student₁—O heck! (The student sits down.)

Teacher—What do you think of that problem?

Student₄—No good. No telephone company would have only twenty subscribers.

Teacher—What's the value of the problem then?

Student₄—Just to give us some idea about calculus.

Teacher—Any questions on this problem?

Student—What would happen if that was 5 cents instead of 10 cents in the telephone problem?

Teacher—That is a very good question. Shall we work it and see what happens?

Student₂—There is only five minutes left.

Teacher—Shall we all guess at it and then someone work it out tonight? (Several students guess and the instructor makes note of their estimates. A student volunteers to solve the problem.)

Teacher—Are there other problems?

Student—I have one. (The student reads a problem asking for the maximum feet of fence to enclose a certain rectangle if there is a cross fence.)

Teacher—Will you write it on the blackboard so we can work it out for tomorrow? (While the students were copying the problem the bell rang.)

After the class recitation the instructor remarked that he felt that they were traveling at a slow pace, but that when he recalled the hatred the students had had for the subject at the beginning of the year, even intelligent questions were a reward for his effort in the course. Not all the instructors using a cultural type of text were as enthusiastic about general mathematics as this instructor; but, with very few exceptions, they expressed themselves as believing that general mathematics was meeting the needs of the terminal type of student far better than the traditional freshman courses in mathematics. As one instructor stated, "At least the kids like it better." And that raises the question, "Do the 'kids' like general mathematics better than traditional mathematics?"

Success as Indicated by the Students' Questionnaire

Students Studying Preparatory Textbooks

A careful study of Questionnaire B[24] and conversations with the students during class visitations revealed that the students enrolled in the preparatory *general* mathematics were as a rule satisfied with the course. They also expressed themselves as having enjoyed their work in high school mathematics. There were, of course, those who did not enjoy certain topics. The subject matter most often mentioned as disagreeable was emphasis of graphs, irrational functions, determinants, velocity paths of projectiles, acceleration problems, trigonometric identities, permutations and combinations, and imaginary numbers. Such comments as the following were few.

The course is not interesting in that it starts with rather simple beginner's mathematics and leads up to more complicated advanced work, giving the development a very formal use.

I did not particularly enjoy the review of high school mathematics.

I have least enjoyed the making of graphs that are made just for the sake of the graph and not related to any problem. I liked the rest of the work.

I did not like the part dealing with graphs with no problems attached.

The section I liked least was that on falling bodies with air resistance.

The majority of the comments expressed approval of the

[24] Appendix D.

course. The differential calculus, solution of triangles, a study of the straight line and the circle seemed to rank high among the favorite topics. The criticisms usually showed approval of the course as a whole; dissatisfaction was generally voiced regarding certain topics only. The following quotations taken from Questionnaire B indicate this attitude.

The complete course to date has been extremely interesting to me. Of course, there were a few parts that I had difficulty with, such as angular velocity and angular acceleration. Most interesting were integration and differentiation.

The course as a whole is a very interesting course. There are a few parts that have not been interesting, such as permutations, combinations, probability and further properties of circular function. The parts that were best were analytical geometry, differentiation, and integration.

A further study of the questionnaire also revealed that these students had found high school mathematics very interesting. The following statements that were typical concerning college general mathematics perhaps could likewise apply to high school mathematics.

I think this is an excellent course.

On the whole I like the whole course.

I have enjoyed the whole course immensely.

Thus conversations with the students studying a preparatory type of general mathematics and a study of their comments on the questionnaire showed that they were in general satisfied with both their high school and their college mathematics courses.

Students Studying Cultural-Preparatory Textbooks

However, this attitude of satisfaction was not as pronounced among the students studying the cultural-preparatory type of general mathematics, either in the conversations with them during class visits or as revealed by their comments on the questionnaire. The following comments of students of the cultural-preparatory mathematics reveal to some extent the undesirable attitude that has been built up in the minds of many students.

My adviser said this would be a good course for a B.A. student, but it has turned out to be just plain old dull hard mathematics.

I have always found mathematics most difficult, and hence have never found it interesting or enjoyable, this year being no exception.

The only part I liked in this course was logarithms and exponents. The other parts were not interesting at all. I think unless you are particularly fond of mathematics you will not gain anything from this course.

I think this course would be very interesting if one cared for mathematics. Otherwise it is not so interesting.

If the course is for the average students, frankly I think it should be abolished.

From my own experience, I would not recommend it unless for a mathematics major.

I have found this course to be of little or no interest in the main, probably because of the fact that I have very little interest in mathematics and less aptitude in the subject. The more mathematics I take the more I hate it.

As soon as the required year for taking mathematics is up, I'll be very glad indeed.

I think mathematics is very good for developing the mind. It trains one to think, plan, and reason things out, but at times it becomes rather dull and annoying when there is no visible means of usefulness in everyday life. The mathematics in this course seems highly irrelevant.

I'm sorry I have to take this course because I dislike mathematics so. I hate doing it and therefore don't get much out of it. I have never liked mathematics at all.

This course tries to cover many phases of mathematics without going into them deeply enough to show their uses or advantages. For instance, calculus is a wide subject and we aren't given a chance to really use our fundamental principles. We get only as far as to memorize elementary principles in each branch—trigonometry, calculus, analytic geometry, and algebra—but can't put it to work.

The subjects are unrelated and make the course jump around too much.

I enjoy some problems that seem to be practicable in my life, but most of them seem to me to be meant for future engineers.

I think this course, if it is designed for those who do not intend to go on with mathematics, is far too detailed and the problems are ones we will never use. I think it would be better to have some practice in the more general types of arithmetic that we would be more likely to encounter in our everyday life.

The course is too much of everything and not enough of any one specific *practical* mathematics.

I don't believe this course is of any importance except to those people who definitely intend using it. (i.e., engineers, architects, mathematics teachers, etc.) I don't think that mathematics should be a required course in college except for students in fields requiring the special concentration.

My work in this course has been way above the average, but I am not enthusiastic about the course.

I have never enjoyed a course in mathematics and this course is no exception.

For me, this course constitutes the only displeasure at college. It takes a great deal of my time and I give it grudgingly.

It doesn't help morale much by being surrounded by mathematics majors who have had four years of high school mathematics and I only had two years.

I believe this course is far too technical for the average student who does not intend to continue with mathematics in future terms. It seems that because we know we will not use our mathematics later on in college and that it is so useless, we dislike it so. I feel mathematics courses should be elective and only those desiring to take them should be burdened with them. Mathematics for me is a true drudgery.

This course is the worst of all the courses I've ever had to take. I hate it because it is boring, of little practical value as far as I am concerned, etc. I believe that this course should be given only to those who are really interested in it and to whom it may be of some advantage as far as the development of their mental power is concerned. As for myself (and others in the class) it is purely a waste of time—and (censored).

The whole course seems to be utterly useless for later life and I see no reason why B.A. students should be forced to suffer through one year of it.

I do not believe that this course is much use for people who major in music, for example.

Mere numbers and especially formulas whose fundamental rational basis is not explained to the student prove most boring and valueless.

I think that this course is of no interest or value to the average student. I never expect to want to know how much mud the Mississippi carries to its delta in .16 of a second.

The course was fairly interesting but I think even a year of bookkeeping would be of more value to the average student than taking this course.

The course not only seems uninteresting but absolutely useless.

There was a group of students among those studying the cultural-preparatory course who seemed to enjoy the course and expressed themselves as believing that it would be of value to them in their future work. These students, in general, were those who expected to enroll in other mathematics courses or expected to take courses requiring a mathematical background. The following comments from the questionnaires of students using the cultural-preparatory textbook illustrate the tendency of the students to express themselves in favor of the course if they feel that they will use it in their future work.

I think that this course gives a good preparation in mathematics which I'll need if I ever get to be an engineer.

I think the course on the whole is fairly interesting for a mathematics course and I must have mathematics in advanced radio work.

On the whole I think mathematics is a very fascinating subject from the point of view of satisfaction given to a student in mastering it. . . . I expect to teach it some day.

The course in general mathematics is the most interesting mathematics course I have ever taken. I think the reason for liking it so much is that a topic is in hand just long enough to create interest and not too long to cause boredom. . . . I have always liked mathematics.

I found this course very interesting on the whole, but I think mathematics majors like any kind of mathematics.

I thoroughly enjoyed this course, as I do any mathematics course.

I have found the work this term interesting and have enjoyed it more than high school mathematics.

I think mathematics is a fairly interesting subject on the whole and I enjoy this mathematics more than any other mathematics course I ever took in high school. I am fond of the study on the basis that it proves something at which you can spend time in actual thought—objective, clear thinking, yet somehow, I usually find myself after the first few weeks in a mathematics class wondering just what my fellow students and the instructor are talking about. I have a great deal of difficulty grasping most of the class work—yet I am very anxious to learn the course.

A study of the questionnaires from the cultural-preparatory group revealed that two thirds of the students found the college mathematics as interesting as high school mathematics and one third found it either more interesting or less interesting than high school mathematics. For every student who indicated that he found college mathematics more interesting than high school mathematics, there were nearly four students of the opposite opinion. Thus the cultural-preparatory mathematics seemed to be losing in favor with the students.

The students were not only willing to indicate their interest in the general mathematics course by checking the questionnaire but they were very cooperative in listing the topics they liked most and those they liked least, with comments and suggestions regarding the course. Typical of such comments are the following:

During both terms I have enjoyed *this* course tremendously, particularly: trigonometric functions and their relations, exponents, logarithms, and the theory of equations, and the differentiation of algebraic functions. There are no parts which I can say truthfully that I did not enjoy.

The parts of this course which I have enjoyed most are: functions and graphs, straight line formulas, logarithms, trigonometry, and differentiation of algebraic functions. The parts of the course which I have enjoyed least are: joint variation, theory of equations, and integration.

I have liked any problems which have a definite practical value—problems consisting of vague suppositions and just numbers seem a waste of time. In other words, I think mathematics should be a course to help us in everyday life rather than a perplexing series of problems. This course seems to have quite a lot of practical application.

The first part of the book interested me more than the last. The parts most interesting were those containing trigonometry, logarithms, and relations among trigonometric functions. I found the parts dealing with differentiation and integration *very* difficult.

I prefer the second part of the course—trigonometry.

In the first section, I found solving of trigonometric functions and logs very interesting but did not like slopes and graphs for there was too much repetition.

I enjoyed working at interpolation and logarithms. The remaining parts in the course were uninteresting and very difficult to understand.

I enjoyed the following parts most: straight line formulas, exponents, logarithms, and trigonometry. I did not enjoy very much the sections about the theory of equations, differentiation, problems dealing with the instantaneous rate of change and falling bodies, etc. However, parts of these divisions I did like—maximum and minimum, derivatives and synthetic division. On the whole I like it.

I think the section on equations is very interesting but I do not like graphs.

The only mathematics in this course that has aroused my interest and brain is that which has practical application.

I disliked intensely Horner's method, differentiation, and integration.

Graphs—very *boring*.

Projectiles—terrible.

Trigonometry and logarithms are interesting but graphs and their problems dull.

From compound interest problems I got no sense of utility.

The problems in the text have too much to do with physics.

I didn't like the problems concerned with throwing things up and down. They still confuse me.

The trigonometry in the course was very interesting. However, I think that the calculus should be taught to students who are able to pick it up quickly, rather than to those who are slow in learning it.

The study of the exponential function was a waste of time.

In this course we delve too deeply into the graph system.

Speeds, acceleration, are useless for me, as I see it.

It is one of the most interesting courses to me because of the practical applications which it gives.

This course, as stated above, has been fairly interesting. I would wish, however, to state that it could have been more interesting if more fundamental ideas of numbers and the theories behind mathematics in general could have been given instead of drill on formulas and practical problems. Mathematics could be a fascinating subject even for one who, like myself, is not proficient in the subject, if some method were devised for more thoughtful work and less memory drill.

The topics listed frequently among the desirable topics were logarithms, differentiation, straight line, solution of triangles, and conic sections and graphs.

Those topics that students disliked were projectile problems, acceleration, rate problems, Horner's method, joint variation, trigonometric identities, integration, logarithms, trigonometry, differentiation, and graphs.

It will be noticed that the last four topics listed as undesirable to many students were also listed as of special interest to many others. This disagreement as to desirable content was also noticeable not only in the comments given on the questionnaire but also in talks with students on visitations of classes. While the students studying the preparatory mathematics were in general agreement concerning the interest in the topics, the cultural-preparatory group was divided as to the topics liked best and those liked least.

Students Studying Cultural Textbooks

It would be well now to look at some of the comments from the students studying the cultural type of mathematics. It is neither possible nor desirable to reproduce all the quotations that appeared on the students' questionnaires; however, it is impossible to get the personal nature of their likes and dislikes except as they are expressed in their own words. Several dissatisfied students expressed their displeasure in such comments as the following:

As far as a mathematics course goes it just isn't, but as social science majors must take mathematics I think this survey is the most adequate kind of a course for them.

I find this course to be of little interest because it is all mixed up. You only scratch the surface and learn just the essential points of mathematics.

I don't think the course has any value. The student cannot get anything valuable out of the course except the part of the work concerned with reasoning (truth or validity). I found the rest of the work boring.

This course holds little interest for me in relation to the mathematics I have heretofore taken. It is true that I found the opening chapter on validity and truth very interesting. I found in that chapter a model type of thinking that all should strive to imitate and, therefore, found in that particular section of the course something constructive. I am sorry to say that the rest of the course does not seem to present to me anything constructive or helpful in my future life.

Frankly, I haven't found this course of much interest. I enjoy working with concrete facts and figures rather than with the logic behind our mathematics. I did find the latter part of the term, which dealt with equations, more interesting than any of the other topics we have taken. I do my best work and am most interested in those courses where more mechanical methods are used. This may be no testimonial of my intelligence, but frankly speaking—it is so. Perhaps if this course was given before our algebra (high school) courses it would be more interesting and of more value. Once you use the mechanical methods so long, it is difficult to go back to the underlying reasons and start anew.

Some of the more popular topics among the students studying the cultural general mathematics textbooks were logic in algebra, logic in geometry, inductive and deductive reasoning, non-Euclidean geometry, basis of number systems, natural numbers, study of the foundations of geometry, relation of mathematics to other fields, historical topics, unsolved problems, simple graphs, and functions. Those topics disliked by students, listed according to the frequency of expression on the questionnaire, are imaginary and irrational numbers, powers and roots, equations of higher degree, analytic geometry especially conic sections, solution of oblique triangles, and integral calculus. Some of the reasons for dissatisfaction and some suggestions for improvement are given in the students' own words.

It should contain more applied mathematics.

The part I've liked least was working with fractions.

I can't see how this mathematics can help me in the future. It can't be used in science or in the armed forces.

The one drawback to this course is the manner of approach. At first it is a novelty, and hard to grasp, for one cannot unlearn what one has spent years learning.

I think the part on unsolved and impossible problems is very interesting.

As far as I'm concerned this course has many values aside from the purely mathematical viewpoint. It helps to develop sound, logical, detailed, and more profound thinking. It helps to explain the philosophical attitudes in science and mathematics. It helps to establish a basis in mathematics which gives you some background to rely on.

Irrational numbers, fractions, and square root are boring.

The most interesting part is the section on how numbers and the number system developed.

I don't think this course is as scientific as a mathematics course should be.

I dislike extracting and obtaining square roots and working with powers and roots in general.

I also enjoyed working with equations.

I have finally found out why I do my mathematics in the way I do.

Mathematics survey course is very interesting because it gives the reasons for solving problems which we took for granted.

I find the part dealing with linear and quadratic equations a bit vague.

I didn't like complex numbers or irrational numbers.

I liked non-Euclidean geometry because it was something different from anything I had ever studied. This topic also simplified many of the difficult theorems studied in plane geometry.

I didn't enjoy analytical geometry because some of the problems contained difficult and long computations.

I dislike working with exponents and the like immensely—probably because I never did understand them.

I did not fully appreciate the work on conic sections and such.

I find this course very interesting because it allows for a deeper insight into the mathematics principles that were previously employed blindly. The most interesting parts of the course I find are the discussions which we have in class.

It probably is the first time I've learned to look at mathematics and realize that it is more than abstract.

I didn't like finding the limit of a number and still don't understand the method used. Integration is also confusing.

I prefer Mathematics Survey because in it the emphasis is on ideas rather than on answers.

The work on motion for which I think I needed some theoretical work in physics became difficult and confusing because of the wording.

I enjoyed most in this course the discussion of Euclidean geometry as compared with Lobachevskian.

Analytic geometry was of very little interest to me.

The study of algebraic sets was quite interesting.

The study of logarithms was of little interest.

The most obscure part of the book was the chapter on limits and the calculus.

The idea of how our number system originated was one phase which held my intense interest.

The logic will help me to look at other matters not concerning mathematics in a new light. It helps me to think more clearly, too.

Those parts of most interest have been the number systems and the introductory phase of the course, such as theories of induction and deduction, etc.

The most interesting parts of the course are those that deal with logic, deductive and inductive, and also with truth and validity.

This course more than any other mathematics course that I have taken helped me to understand the logic behind the mechanical application I have been using in high school.

The parts which in my opinion are least enjoyable and unnecessary are the various laws of addition, etc. They are completely useless and meaningless. They do not serve any purpose or logic.

This logic has already come in very handy for me in judging opinions and decisions logically.

I think the system of notation is one of the most interesting parts of the course.

I don't particularly enjoy factoring.

Strong prejudice against mathematics has prevailed so long with many students that favorable comments came as a sort of secret confession. These comments concerning their rebellion against mathematics and conversion to a spirit of mathematical appreciation do not, because of their very nature, lend themselves readily to succulent paraphrase without obscuring the view of the personal reactions. Permit me, therefore, to quote again directly from the students:

Very interesting course—one of the few mathematics courses that I have ever taken which I enjoyed.

Although this course is general, I think that it is good for the student who has never really understood what he was doing in mathematics.

Very good! As a student who is majoring in the arts and who will have no need for the involved and technical branches of mathematics, I would have pined for this course if it were not offered.

On the whole, considering that mathematics is my most unpopular subject, I think that this is about the most interesting mathematics course I have ever taken.

When I start a topic I don't like it but when I'm through it is fairly interesting.

The most interesting part of this course is the method of teaching it. It is an approach that assumes (and rightly so) that the student is a rational human being and, as such, is to be taught logical relationships and *not* mechanical contrivances. I like the general "debunking" of the machinelike, unimaginative, "unorganic" method used in high schools to teach mathematics. In the sense that it is taught now mathematics is an active not a passive force. I think that if this method was applied to higher mathematics it would probably be a fascinating study.

Although this course had led me into confusion in many instances, it has given me an insight into the meaning and purpose of mathematics which I never before saw. A very valuable course. My confusion is the result of previous mathematics experience.

I think this course is very interesting because: (1) In no other course has the reason for mathematics been described, nor the history of mathematics touched upon. (2) The actual process of working the problem was secondary to the reason why it existed and the method of treating it.

This course is excellent because it gives students who cannot or do not want to continue in mathematics an acquaintance with this subject. This cultural aspect might not be too important but I feel in a liberal arts course "Survey Mathematics" is really a worthwhile subject.

I feel that this course has been very beneficial to me as a college student who wanted to find a general interest in the subject. By exploring various fields of mathematics I have met up with various theories of our universe and in general have been introduced to new fields which broadened my views.

While some parts of the course still are not quite fully grasped (Lobachevskian geometry, for instance), I have found this course very interesting. Furthermore, I am amazed to find that I have taken a liking to a subject which formerly held very little interest for me. In fact, I detested some of the mathematics taught in high school.

This is a remarkably fine course for those who desire merely a general and fairly comprehensive perspective of all mathematics. But it has little place for those who have a need for specified mathematics.

To sum up my opinion, this course is a very interesting survey of mathematics rather than a practical mathematics course.

I find the historical element in this course very interesting.

I am not an excellent pupil in this mathematics course but I can truthfully state that I thoroughly enjoy it.

I think that this is the ideal course to wind up my mathematics career with, as I now learn *why* I did *what* I did in high school.

As a matter of fact though, on the whole, I steer clear of "mathematics." I do enjoy it when skill with numbers is not too important.

I also like the idea of not forcing Arts majors to take specialized courses in mathematics which will be of no use to them later on. This course gives the student a general background in mathematics which is sufficiently detailed to serve his purposes.

I found this course much more interesting than any of the mathematics courses I took in high school.

As a student who expects to major in a subject that will have little in common with mathematics, and wishes to find in her single year of college mathematics an idea of all pure mathematics and its applications, I find the course very valuable.

I think this course should be taken by those who will not use mathematics in future years.

This is the only truly interesting mathematics course that I have taken.

This course has been a constant source of interest to me! When I assigned myself to this course I didn't think that it was going to be interesting—and much more, practical. However, I can say my "pre-conceived" ideas were entirely wrong. Of course, it may be the instructor who has made the course so interesting, but in this case I think the excellent instructor just helped to bring out the course.

This is the first time I have ever really enjoyed *any* mathematics course.

I firmly believe that Survey Mathematics answers the hackneyed question, "What's the use of my taking mathematics at all?" As far back as I can remember I would tearfully stand before my family holding a report card whose red circled face loudly proclaimed—D in Mathematics. It was at these times that I would invariably ask the aforementioned question. My uncle, a professional mathematician, would always answer, "It teaches you to think." I know now why I never understood him. That mathematics never did teach me how to think—but Survey Mathematics has. I've learned to delve into problems, and to analyze certain questions which came before me. I can comprehend for the first time why some people are fascinated by mathematics. In fact, I've really given mathematics a fresh chance.

All through my schooling I have had difficulty in mastering the phases of mathematics, recollecting even to my public school days when my teacher hit me on the head with a blackboard eraser for not understanding long division. But as I grew older I have found mathematics to be a highly fascinating course, to be wondered at. I really think that, if nothing else, I have gotten out of this course a common basis for discussion with other college students. I'm almost sure that advanced mathematics will be of little value to my future profession in life (social work) but I have honestly been intrigued all through this course.

It is well to observe at this point that these students enrolled in the cultural mathematics course were on the whole as dissatisfied with mathematics when they enrolled in the course as the group in the cultural-preparatory course. In fact, in rating high school mathematics on the questionnaire, 27 per cent of the cultural group checked "very interesting," 61 per cent "fairly interesting," and 12 per cent, "of little interest," whereas the cultural-preparatory group checked the same items according to the following percentages: "very interesting" (30 per cent), "fairly interesting" (60 per cent), "of little interest" (10 per cent). These proportions indicate that the cultural-preparatory group were at least as favorably impressed with their high school mathematics as the cultural group.

However, at the close of the freshman year of college, one third of the preparatory mathematics students had changed in their reaction toward mathematics, according to the questionnaire, but for every one who was more favorable to mathematics there were four who had become less interested in the subject.

On the other hand, two thirds of the students using a cultural textbook indicated that their interest in mathematics had changed and nearly 60 per cent of these students indicated that general mathematics was more interesting than the traditional mathematics of high school.

Summary—Success in Meeting the Objectives of General Mathematics

Thus it seems from an examination of the comments of more than fifteen hundred students enrolled in general mathematics that those students who expect to use mathematics as a tool subject are, in general, satisfied that the course is meeting their needs. In the large academic group studying the *cultural-preparatory* general mathematics there is dissatisfaction with the course and disagreement as to the cause of dissatisfaction. The purposes and aims of this group of students, according to their statements, are not adequately met. The cultural group is more nearly in harmony concerning the suggestions for improving the course, and, while all are not satisfied, the general feeling is expressed that general mathematics is more interesting than the traditional high school mathematics. In general, the opinions of the students as expressed on the questionnaire also are in harmony with the judgment of the instructors as indicated both by the comments on the questionnaires and by statements made during interviews.

CHAPTER VII

SUMMARY AND CONCLUSION

Purpose of Study

WHAT are the objectives of mathematics that have been proposed for students who are not majors in that field? What content material is being offered? What success is being attained in meeting the proposed objectives of these courses? It was these and other closely related questions that gave impetus to this investigation.

In Chapter I the aims of the study are stated formally to be:

1. To trace the historical development of college general mathematics in the United States.

2. To show the present status of general mathematics in American colleges.

3. To discover and point out certain trends in the development of college general mathematics.

Reports of Other Investigators

The reports of other investigators in the field show that various methods have been used to determine the content of the general mathematics course. Among these methods of selection of content are (a) analysis of the felt needs of individuals in the field, (b) survey of mathematical requirements of students as indicated by a study of the textbooks in the social sciences, natural sciences, and the calculus, and (c) study of the opinions of specialists in the field. Several investigations to determine the mathematical offerings in the freshman year were reported. These studies were either restricted in scope or based upon an analysis of the catalogues of certain colleges. Although this is valuable research, as Eells has pointed out, there is a difference between the offerings of a college and the courses actually given. Little information was reported concerning general mathematics courses that are really given in the freshman year. The attempt to evaluate these courses in terms of their objectives has been slight.

Rise of General Mathematics Movement

In the early development of mathematics no attempt was made to separate mathematics into its various branches until the beginning of the eighteenth century. As specialization continued the separate courses in mathematics became better defined and more uniform until they were called "water-tight compartments" at the beginning of the twentieth century. Some of the many factors that contributed directly to the development of general mathematics for college freshmen are:

1. A new psychology of learning and philosophy of education, which was not in harmony with some of the objectives of traditional mathematics, was being accepted by many educators.

2. Because of the low degree of mastery exhibited by the average student and the high rate of mortality of freshmen in college mathematics classes, there was a growing dissatisfaction with the organization and teaching of freshman mathematics.

3. To re-evaluate and reorganize the curriculum was popular during the early part of the twentieth century.

4. The increase in college enrollment caused both a wide variation in the objectives of the students and an increase in the number of mediocre students.

5. The rapid growth of the junior college and its acceptance as a part of the secondary school was causing a reorganization of the curriculum of the freshman and sophomore years in harmony with the aims of general education.

6. The general mathematics movement in high school was being accepted as a forward step in the development of a mathematics course for the average student.

7. The frontier thinkers in mathematics education were advocating the need for the revision of the mathematics curriculum. This need for reorganization was expressed by John Perry, Felix Klein, E. H. Moore, the report of the Committee of Ten, the report of the National Committee of Mathematics Require-

ments in 1923, the report of the Joint Commission, and the Progressive Education report of 1940.

Thus, to give a perspective and to serve as a frame of reference from which to examine the objectives of general mathematics, these contributing factors effecting the rise of the general mathematics movement have been described in Chapter III.

Objectives of General Mathematics

In Chapter IV the aims of mathematics education have been shown to be in harmony with and complementary to the objectives of general education. The objectives of general mathematics as indicated by (a) committees of specialists in the field, (b) authors of the textbooks, and (c) teachers of the subject, both by their statements and by demonstrations in the classroom, fall into three categories. The purposes of one group are preparatory in nature; the aims of another group concern the cultural, social development of the individual; in the third group the objectives include a combination of the cultural and preparatory functions.

Provisions for Meeting the Objectives of General Mathematics

In investigating the provisions to meet these objectives the investigator found that in both the preparatory and the cultural-preparatory courses the content, style of presentation, and emphasis given to topics were substantially identical, but these differed markedly from the offerings in the cultural type of general mathematics.

Success in Meeting the Objectives of General Mathematics

Data concerning the achievement of the objectives of general mathematics are presented in Chapter VI. In general the theme of this chapter was that (a) the objectives of the preparatory type of mathematics are to a large degree being realized, (b) the realization of the dual aims and purposes of the cultural-preparatory are being seriously questioned, (c) cultural general mathematics, while not entirely satisfactory, is more nearly meeting the needs of the terminal student in mathematics than the traditional offerings.

It was found that the greatest dissatisfaction among the instructors of the cultural-preparatory mathematics was among those whose objectives were of a cultural nature or a combination of cultural and preparatory. The view that the student will derive great cultural value from the study of abstract mathematics is not a new one. In 1911 the American Committee of the International Commission on the Teaching of Mathematics found that the first aim of mathematics as stated by the teachers of mathematics was "the development of analytical power and logical keenness." [1] The Committee further reports that the teachers attempt to develop certain powers of the student, such as "mathematics dexterity in analytical trigonometry and parts of the calculus" and "an analytical power in analytical geometry."

This practice of requiring the terminal mathematics student to pursue the same courses in the freshman year as the mathematics major has been shown in Chapter VI to be unsatisfactory in the opinion of many teachers and students. It has also been condemned by leaders in the field of mathematics education,[2] and as one has said: "We need to draw a sharp distinction today between the place of mathematics in the education of the general reader and the place of mathematics in the education of persons who plan to undertake systematic and rigorous study of science and mathematics in later years." [3]

The Joint Commission clearly takes the stand "different mathematical programs are needed." They report ". . . it may

[1] International Commission on the Teaching of Mathematics, American Report Committee, No. X, Undergraduate Work in Mathematics in Colleges of Liberal Arts and Universities, p. 18. U. S. Bureau of Education, Washington, D. C., 1911.
[2] C. H. Butler and F. L. Wren, *The Teaching of Secondary Mathematics*, pp. 82-84. McGraw-Hill Book Company, Inc., New York, 1941.
Also A. C. Rosander, "Quantitative Thinking on the Secondary School Level." *The Mathematics Teacher*, XXIX, pp. 51-66, 1936.
[3] Raleigh Schorling, "The Place of Mathematics in General Education." *School Science and Mathematics*, XL, No. 1, p. 14, 1940.

have been taken for granted that all students who studied mathematics should be given the same material. . . . The validity of these assumptions is open to serious question. . . . different groups of college students, with different purposes, have different types of mathematical needs. Non-specialists, particularly those having a general or liberal arts interest, need very little command of the higher techniques; but they do need the sort of insight and appreciation just mentioned—in short, familiarity with the cultural significance of mathematics." [4]

Wren states, ". . . in spite of theoretical pronouncements . . . relatively little has been done thus far to differentiate the mathematics courses . . ." [5]

The data in Chapter VI have shown that there is an attempt to meet the needs of this large academic group of non-mathematics specialists and the attempt is meeting with some degree of success in the opinion of both the students and the teachers.

CRYSTAL GAZING

While it is always unsafe to predict the changes that will take place in any of the great fields of learning—and especially is this true during a national crisis—yet it should be observed that in the field of freshman mathematics at present at least the preparatory type of general mathematics has reached the most stable position.

"The customary mathematical program for these preparatory students is in general adequate and suitable to their needs." [6] This statement is substantiated by the data in Chapter VI. It should be observed that the preparatory general mathematics has endured the test of the classroom longer than either the cultural-preparatory or the cultural, according to the reports of the teachers on the questionnaires. In fact, two thirds of the

---

[4] The Final Report of the Joint Commission of the Mathematical Association of America and the National Council of Teachers of Mathematics, *The Place of Mathematics in Secondary Education*, pp. 154, 155. Bureau of Publications, Teachers College, Columbia University, New York, 1940.

[5] C. H. Butler and F. L. Wren, *The Teaching of Secondary Mathematics*, p. 82. McGraw-Hill Book Company, Inc., New York, 1941.

[6] *Ibid.*

teachers using a preparatory general mathematics textbook reported that the course had been offered in their institution more than five years, while less than one third of the teachers using the cultural general mathematics indicated that it had been offered as long as five years in their schools. The trend toward cultural mathematics is further indicated by the fact that forty per cent of the teachers of these cultural courses indicated that they had offered the course three years or less.

This tendency of the preparatory mathematics to become static while the cultural mathematics is becoming more experimental is also shown in the questionnaires from institutions that were either discontinuing or organizing courses in general mathematics. Exactly the same number of schools reported the dropping of general mathematics from the curriculum as those who reported that they expected to offer such a course the following year, but three fourths of those schools who were discontinuing general mathematics indicated that they had been using a cultural-preparatory textbook.

Although only a little more than half the schools that expect to offer the course the following year indicated the textbook they expected to use, of those that did, nearly 80 per cent indicated a cultural general mathematics text.

Although it may seem desirable that this trend of cultural mathematics for the terminal student should increase—many instructors interviewed expressed themselves as favorable to such a program—on the other hand these same instructors have issued the warning that the national crisis may cause students to elect the traditional courses. In fact the present enrollments in a few colleges indicate that because some students desire to prepare for opportunities in the armed forces, there is an increase in the enrollment of the traditional courses and a decrease in enrollment in the general mathematics classes. However, many of the instructors interviewed feel confident that after the national emergency the movement for cultural mathematics will be renewed with great emphasis.

"It is true that the unrest and the dissatisfaction with the teaching of mathematics in the schools and the colleges, which are making themselves felt with increasing strength throughout the land, confront us with a serious and difficult problem." [7] But it is also equally true that the attempts that are being made to solve these problems have met with at least partial success.

Although the opinions of the students as given in Chapter VI indicate that the terminal mathematics student is more interested in the cultural type of course, it is not advocated that the mathematics curriculum should be organized on the sole basis of the interests of the student. Yet it seems evident that more effective teaching will take place if that interest is secured. As Dewey has counseled, teachers should be "more concerned about creating a certain mental attitude than they are about purveying a fixed body of information." [8]

It might be well for even those who "do not care what the student thinks" to recall these words of admonition: "We take pride in the brilliant record which our younger mathematicians are making in research, but to support this superstructure we must create among the intelligent public a broader interest in mathematics and a better appreciation of its values. An important factor here will be the attitude of the college-trained man in the community and this will become more tolerant or even enthusiastic if we can improve the quality of the undergraduate instruction." [9]

[7] A. Dresden, "Program for Mathematics." *American Mathematical Monthly,* Vol. 42, p. 199, April, 1935.

[8] John Dewey, "The Supreme Intellectual Obligation." *Science,* Vol. 79, p. 242, March 16, 1934.

[9] J. I. Tracey, "Undergraduate Instruction in Mathematics." *American Mathematical Monthly,* Vol. 44, p. 288, May, 1937.

BIBLIOGRAPHY

I. GENERAL MATHEMATICS TEXTBOOKS

ANDERSON, W. E. *A First Course in College Mathematics.* Harper and Brothers, New York, 1933.

BRESLICH, E. R. *Correlated Mathematics for Junior Colleges.* The University of Chicago Press, Chicago, 1928.

BRINK, R. W. *A First Year of College Mathematics.* D. Appleton-Century Company, Inc., New York, 1937.

BRUCE, R. E. *A Survey of Elementary Mathematics.* Boston University Book Stores, Boston, 1936.

CLAWSON, JOHN W. *Freshman Mathematics.* (Planographed by) Edwards Brothers, Inc., Ann Arbor, Mich., 1937.

COOLEY, H. R., GANS, D., KLINE, M., AND WAHLERT, H. E. *Introduction to Mathematics.* Houghton Mifflin Company, Boston, 1937.

CORNELL UNIVERSITY, DEPARTMENT OF MATHEMATICS. *Elementary Concepts of Mathematics.* Edwards Brothers, Inc., Ann Arbor, Mich., 1940.

CURRIER, C. H. AND WATSON, E. E. *A Course in General Mathematics.* The Macmillan Company, New York, 1929.

CURRIER, C. H., WATSON, E. E. AND FRAME, J. S. *A Course in General Mathematics.* The Macmillan Company, New York, 1939.

DARKOW, M. D. AND LANDERS, M. K. *Elementary Mathematics.* Edwards Brothers, Inc., Ann Arbor, Mich., 1941.

DAVIS, H. T. *A Course in General Mathematics.* The Principia Press, Bloomington, Ind., 1935.

DRESDEN, A. *Invitation to Mathematics.* Henry Holt and Company, New York, 1936.

ELLIOTT, W. W. AND MILES, E. R. C. *College Mathematics—A First Course.* Prentice-Hall, Inc., New York, 1940.

GALE, A. S. AND WATKEYS, C. W. *Elementary Functions and Applications.* Henry Holt and Company, New York, 1920.

GEORGES, J. S. AND KINNEY, J. M. *Introductory Mathematical Analysis.* The Macmillan Company, New York, 1938.

GRIFFIN, F. L. *Introduction to Mathematical Analysis.* Houghton Mifflin Company, Boston, 1936.

HARKIN, D. *Fundamental Mathematics.* Prentice-Hall, Inc., New York, 1941.

HELLIWELL, C. H., TILLEY, A. AND WAHLERT, H. E. *Fundamentals of College Mathematics.* The Macmillan Company, New York, 1935.

HILL, M. A., JR. AND LINKER, J. B. *Introduction to College Mathematics.* Henry Holt and Company, New York, 1938.

HUNTER COLLEGE OF THE CITY OF NEW YORK, DEPARTMENT OF MATHEMATICS. *General Mathematics.* (Lithographed by) Polygraphic Company of America, Inc., New York, 1938.

JOHNSTON, F. E. *Introductory College Mathematics.* Farrar & Rinehart, Inc., New York, 1940.

KARPINSKI, L. C., BENEDICT, HARRY Y. AND CALHOUN, JOHN W. *Unified Mathematics.* D. C. Heath and Co., New York, 1918.

KASNER, E. AND NEWMAN, J. *Mathematics and the Imagination.* Simon and Schuster, New York, 1940.

KOKOMOOR, F. W. *Mathematics in Human Affairs.* Prentice-Hall, Inc., New York, 1942.

LASLEY, J. W. AND BROWNE, E. T. *Introductory Mathematics.* McGraw-Hill Book Company, Inc., New York, 1933.

LENNES, N. J. *A Survey Course in Mathematics.* Harper and Brothers, New York, 1926.

LOGSDON, M. I. *Elementary Mathematical Analysis,* Vol. I. McGraw-Hill Book Company, Inc., New York, 1932.

LOGSDON, M. I. *Elementary Mathematical Analysis,* Vol. II. McGraw-Hill Book Company, Inc., New York, 1933.

LOGSDON, M. I. *A Mathematician Explains.* The University of Chicago Press, Chicago, 1935.

MACKIE, E. L. AND HOYLE, V. A. *Elementary College Mathematics.* Ginn and Company, Boston, 1940.

MERRIMAN, G. M. *To Discover Mathematics.* John Wiley and Sons, Inc., New York, 1942.

MILES, H. J. *A First Year College Mathematics.* John Wiley and Sons, Inc., New York, 1941.

MILNE, W. E. AND DAVIS, D. R. *Introductory College Mathematics.* Ginn and Company, Boston, 1935.

MINSSEN, H. F. AND MEYERS, W. H. *Survey of Mathematics.* San Jose State College, San Jose, Calif., 1940.

MOORE, JUSTIN H. AND MIRA, JULIO A. *The Gist of Mathematics.* Prentice-Hall, Inc., New York, 1942.

MORTON, R. L. *Teaching Arithmetic in the Intermediate Grades.* Silver Burdett Company, New York, 1939.

MULLINS, G. W. AND SMITH, D. E. *Freshman Mathematics.* Ginn. and Company, Boston, 1927.

MUNSHOWER, C. W. AND WARDWELL, J. F. *Basic College Mathematics.* Henry Holt and Company, New York, 1942.

NEUREITER, P. R. *A Workbook in General Mathematics.* Burgess Publishing Co., Minneapolis, Minn., 1940.

NEWSOM, C. V. *An Introduction to Mathematics.* (2 vols.) The University of New Mexico Press, Albuquerque, N. M., 1940.

PHILIP, M. *Mathematical Analysis.* Longmans, Green and Co., New York, 1936.

RANSOM, W. R. *Freshman Mathematics.* Longmans, Green and Co., New York, 1925.

RICHARDSON, M. *Fundamentals of Mathematics.* The Macmillan Company, New York, 1941.

RICHTMYER, C. C. AND FOUST, J. W. *First Year College Mathematics.* F. S. Crofts & Co., New York, 1942.

ROWE, J. E. *Introductory Mathematics.* Prentice-Hall, Inc., New York, 1927.

SLICHTER, C. S. *Elementary Mathematical Analysis.* McGraw-Hill Book Company, Inc., New York, 1918.

SLOBIN, H. L. AND WILBUR, W. E. *Freshman Mathematics.* Farrar & Rinehart, Inc., New York, 1938.

STEPHENS, R. P., BARROW, D. F. AND BECKWITH, W. S. *Freshman Mathematics.* Division of Publications, The University of Georgia, Athens, Georgia, 1936.

THEOBOLD, REV. JOHN A. *Freshman Mathematics.* Edwards Bros., Inc., Ann Arbor, Mich., 1941.

UNDERWOOD, R. S. AND SPARKS, F. W. *Living Mathematics.* McGraw-Hill Book Company, Inc., New York, 1940.

WELLS, V. H. *First Year College Mathematics.* D. Van Nostrand Company, Inc., New York, 1937.

WOODS, F. S. AND BAILEY, F. H. *A Course in Mathematics.* Ginn and Company, Boston, 1907.

YOUNG, J. W. AND MORGAN, F. M. *Elementary Mathematical Analysis.* The Macmillan Company, New York, 1917.

ZANT, J. H. AND DIAMOND, A. H. *Elementary Mathematical Concepts.* Burgess Publishing Company, Minneapolis, 1941.

II. MAGAZINE ARTICLES

ADAMS, L. J. "Mathematics in California Junior Colleges." *Junior College Journal,* Vol. 7, pp. 194–195, January, 1937.

BARNETT, J. "A Proposal for the Improvement of the Teaching of Mathematics." *National Mathematics Magazine,* pp. 74–81, December, 1934.

BLEDSOE, J. "Failures in College Freshman Mathematics." *Texas Outlook,* Vol. 24, pp. 18–20, October, 1940.

CAMERON, E. "Program in Freshman Mathematics Designed to Care for a Wide Variation in Student Ability." *American Mathematical Monthly,* Vol. 47, pp. 471–473, August, 1940.

CAMPBELL, ALAN D. "Some Mathematical Shortcomings of College Freshmen." *The Mathematics Teacher*, Vol. 27, pp. 420–425, December, 1934.

DORWART, H. "Comments on the North Carolina Program in Freshman Mathematics." *American Mathematical Monthly*, pp. 37–39, January, 1941.

DRESDEN, A. "Program for Mathematics." *American Mathematical Monthly*, Vol. 42, pp. 198–208, April, 1935.

GINNINGS, P. "Mathematics and Science Requirements for the Liberal Arts Degree in Southern Colleges." *High School Quarterly*, Vol. 23, pp. 10–12, October, 1934.

GEORGES, J. "Mathematics in the Junior College." *School Science and Mathematics*, Vol. 37, pp. 302–316, March, 1937.

GEORGES, J. "Humanizing the Curriculum of the Natural Sciences and Mathematics." *School Science and Mathematics*, Vol. 40, pp. 454–455, May, 1940.

GRIFFIN, F. "An Experiment in Correlating Freshman Mathematics." *American Mathematical Monthly*, Vol. 22, pp. 325–330, December, 1915.

GUGGENBUHL, L. "Failure in Required Mathematics at Hunter College." *The Mathematics Teacher*, Vol. 30, pp. 68–75, February, 1937.

HILLS, J. "Junior College Mathematics." *School Science and Mathematics*, Vol. 29, pp. 880–885, November, 1929.

HOLROYD, I. "Weakness in High School Students Who Enter College Mathematics and a Suggested Remedy." *The Mathematics Teacher*, Vol. 27, pp. 128–137, March, 1934.

LONG, E. "Correlation of Algebra, Geometry and Physics." *Educational Review*, Vol. 24, p. 309, October, 1902.

O'QUINN, R. "Status and Trends of Ability Grouping in the State Universities." *The Mathematics Teacher*, Vol. 38, pp. 213–215, May, 1940.

PRESSEY, L. "The Needs of Freshmen in the Field of Mathematics." *School Science and Mathematics*, Vol. 29, pp. 238–243, March 30, 1930.

REEVE, W. D. "The Case for General Mathematics." *The Mathematics Teacher*, Vol. XV, No. 7, November, 1922.

RICHTMYER, C. C. "Functional Mathematics Needs of Teachers." *Journal of Experimental Education*, Vol. 6, pp. 396–398, June, 1938.

SANFORD, V. "Textbooks in Unified Mathematics for College Freshmen." *The Mathematics Teacher*, Vol. 16, pp. 206–214, April, 1923.

SCHORLING, R. "The Place of Mathematics in General Education." *School Science and Mathematics*, Vol. 40, No. 1, p. 14, January, 1940.

TRACEY, J. "Undergraduate Instruction in Mathematics." *American Mathematical Monthly*, Vol. 44, pp. 284–288, May, 1937.

TRAINOR, J. C. "New Approach for a Course in Mathematics for Teachers." *School and Society*, Vol. 41, p. 398, March 23, 1935.

WATSON, E. E. "An Analysis of Freshman College Mathematics." *Education*, Vol. 48, pp. 225–228, December, 1927.

WILSON, T. H. "The First Four Years of Junior College." *Junior College Journal,* Vol. 9, p. 365, April, 1939.

YOUNG, J. W. "The Organization of College Courses in Mathematics for Freshmen." *American Mathematical Monthly,* Vol. 30, pp. 4–16, January, 1923.

III. GENERAL REFERENCE BOOKS

ANDERSON, FRANK. *The General Mathematics Course in Higher Institutions.* Unpublished Master's Thesis, University of Arizona, 1938.

BUTLER, C. H. AND WREN, F. L. *The Teaching of Secondary Mathematics.* McGraw-Hill Book Company, Inc., New York, 1941.

CAMPBELL, D. S. *A Critical Study of the Purposes of the Junior College.* George Peabody College for Teachers, Nashville, Tenn., 1930.

COLE, LUELLA. *The Background of College Teaching.* Farrar and Rinehart, Inc., New York, 1940.

COLVERT, C. C. *The Public Junior College Curriculum.* Louisiana State University Press, University, La., 1939.

EELS, W. C. *The Junior College.* Houghton Mifflin Company, Boston, 1931.

EELS, W. C. *Why Junior College Terminal Education?* American Association of Junior Colleges, Washington, D. C., 1941.

EELS, W. C. *Present Status of the Junior College Terminal Education.* American Association of Junior Colleges, Washington, D. C., 1941.

HANNELLY, R. J. *Mathematics in the Junior College.* Unpublished Doctor's Dissertation, University of Colorado, Boulder, Colo., 1939.

HOPKINS, L. T. *Integration—Its Meaning and Application.* D. Appleton-Century Company, Inc., New York, 1937.

JOINT COMMISSION OF THE MATHEMATICAL ASSOCIATION OF AMERICA AND THE NATIONAL COUNCIL OF TEACHERS OF MATHEMATICS. *The Place of Mathematics in Secondary Education.* Bureau of Publications, Teachers College, Columbia University, New York, 1940.

KOOS, L. V. *The Junior College Movement.* Ginn and Company, Boston, 1925.

MCCORMICK, C. *The Teaching of General Mathematics in the Secondary Schools of the United States.* Bureau of Publications, Teachers College, Columbia University, New York, 1929.

NATIONAL COMMISSION ON COOPERATIVE CURRICULUM PLANNING. *The Subject Fields in General Education.* D. Appleton-Century Company, New York, 1941.

NATIONAL COMMITTEE ON MATHEMATICS REQUIREMENTS UNDER THE AUSPICES OF THE MATHEMATICS ASSOCIATION OF AMERICA. *The Reorganization of Mathematics in Secondary Education.* U. S. Bureau of Education Bulletin No. 32, 1923. (1923 Report.)

PERRY, JOHN AND OTHERS. *Discussion on the Teaching of Mathematics.* The Macmillan Company, New York, 1901.

PROGRESSIVE EDUCATION ASSOCIATION. *Mathematics in General Education.* D. Appleton-Century Company, New York, 1940.

SCHAAF, W. D. *A Bibliography of Mathematics Education.* Stevinus Press, Forest Hills, N. Y., 1941.

SCOTT, P. C. *A Comparative Study of Achievement in College Freshman Mathematics.* Unpublished Doctor's Dissertation, George Peabody College for Teachers, Nashville, Tenn., 1930.

SEASHORE, CARL E. *The Junior College Movement.* Henry Holt and Company, New York, 1940.

SEIDLIN, JOSEPH. *A Critical Study of the Teaching of Elementary College Mathematics.* Bureau of Publications, Teachers College, Columbia University, New York, 1931.

STRANG, R. *Personal Development and Guidance in College and Secondary Schools.* Harper and Brothers, New York, 1934.

WHITNEY, F. L. *The Junior College in America.* Colorado State Teachers College, Greeley, 1928.

WRIGHT, H. A. *An Evaluation of Certain Textbooks in General Mathematics for College Freshmen with a View to Formulating a Course which Affords More Satisfactory Preparation for Calculus.* Unpublished Doctor's Dissertation, New York University, New York, 1932.

IV. BULLETINS

COULTER, J. M. *Practical Education.* An address before the students of Indiana University. Carlon and Hollandback, Printers, 1891.

DAVIS, R. R. AND DOUGLAS, H. R. *The Relative Effectiveness of Lecture-Recitation and Supervised-Individual Methods in Teaching Unified Mathematics in College.* University of Oregon Publication, Eugene, 1929.

HANNELLY, ROBERT J. *Mathematics in the Junior College.* The University of Colorado Studies. Abstracts of Theses for Higher Degrees, Boulder, 1939.

INTERNATIONAL COMMISSION ON THE TEACHING OF MATHEMATICS. *Graduate Work in Mathematics in Universities and in Other Institutions of Like Grade in the United States.* The American Report, Committee No. XII (Bulletin No. 6, 1911), Government Printing Office, Washington, D. C., 1911.

INTERNATIONAL COMMISSION ON THE TEACHING OF MATHEMATICS. *Mathematics in the Technical Schools of Collegiate Grade in the United States.* The American Report, Committee No. IX. U. S. Bureau of Education, (Bulletin No. 9, 1911). Government Printing Office, Washington, D. C., 1911.

INTERNATIONAL COMMISSION ON THE TEACHING OF MATHEMATICS. *Undergraduate Work in Mathematics in Colleges of Liberal Arts and Universities.* The American Report, Committee No. X (Bulletin No. 7, 1911). Government Printing Office, Washington, D. C.

MacLEAN, M. S. AND STAFF OF THE GENERAL COLLEGE. *Curriculum Making in the General College.* University of Minnesota, June, 1940.

McCLURE, J. *Failures in Freshman Mathematics.* Read before the University School Conference—Vanderbilt University, May, 1903. *Vanderbilt University Quarterly,* Vol. III, No. 4, October, 1903.

McDOWELL, F. M. *The Junior College,* U. S. Bureau of Education (Bulletin No. 35). Washington, D. C., 1919.

MOORE, E. H. "On the Foundations of Mathematics." *American Mathematics Society Bulletin,* No. X, Vol. 10, pp. 402–424, 1903.

Report of the American Commissioners of the International Commission on the Teaching of Mathematics. (Bulletin No. 14, 1912). Government Printing Office, Washington, D. C.

Report of the Committee on Secondary School Studies Appointed at the Meeting of the National Education Association, July 9, 1892. Government Printing Office, Washington, D. C., 1893.

SCOTT, P. C. *An Abstract of a Comparative Study of Achievement in College Freshman Mathematics.* George Peabody College for Teachers, Nashville, Tenn., 1939.

UPTON, C. B. "The Training of Teachers of Mathematics in Professional Schools of Collegiate Grade." Reprint from *Educational Review,* New York, April, 1911. Report submitted to the International Commission on the Teaching of Mathematics, by Subcommittee (C. B. Upton, Chairman).

WOLFE, JACK. "Mathematics Skills of College Freshmen in Topics Prerequisite to Trigonometry." *New York City, Thirty-ninth Annual Report of Superintendent of Schools, 1936–1937.* 1937.

ZOOK, G. F. *National Conference on Junior Colleges, 1920.* (Bulletin No. 19, 1922). Bureau of Education, Washington, D. C.

APPENDICES

APPENDIX A

LETTER A

(First request sent to 492 junior colleges and colleges for information concerning general mathematics)

In connection with a survey to determine the present status of general mathematics in American Colleges, it is necessary to have the information indicated on the enclosed blank.

I know of no way to secure this information except directly from the source—you who are teaching the subject. It will require only a few minutes of your time, but it is very vital to the study, which I hope will make a contribution to the field of mathematics education.

Although all the information requested on the form is important to the survey, and all will be held in strict confidence (only the results of the entire survey will be used), if there are certain questions that you prefer not to answer, draw a line through these items. But do, please, fill in the remainder of the form *now*, before it is laid aside and forgotten.

A self-addressed stamped envelope is enclosed for your convenience in returning the form.

I greatly appreciate your cooperation, and any comments concerning the study will be most welcome.

APPENDIX B

QUESTIONNAIRE A

SURVEY OF THE STATUS OF GENERAL MATHEMATICS IN AMERICAN COLLEGES

1. Name of text used in General Mathematics
....
2. Author of text ....
3. Number of semester hours' credit received for the course ....
4. Was the course designed primarily for freshmen, sophomores, juniors, or seniors? ....
5. Do you use a mimeographed outline, or syllabus? ....
6. Approximate number of students enrolled in the course during the

entire calendar year ....

7. Number of years a course in general mathematics has been offered in your institution ....

8. Type of student to whom course is offered ....
                              (dull normal, normal, above normal)

9. Underline the phrase or combination of phrases that best describes the course:

   a. Terminal course primarily for a cultural education.
   b. Terminal course primarily for a practical mathematics education.
   c. Preparatory course for further study in mathematics.

10. Is the entire material of the text covered in the course? ....

11. Are there certain chapters that receive little emphasis or are omitted in your course? .... If so, please circle the numbers representing the numbers of such chapters.

    1 2 3 4 5 6 7 8 9 10 11 12 13 14 15 16 17 18 19 20
    21 22 23 24 25 26 27 28 29 30

12. Are there certain chapters that are given special emphasis?....... If so, please circle the numbers representing the numbers of such chapters.

    1 2 3 4 5 6 7 8 9 10 11 12 13 14 15 16 17 18 19 20
    21 22 23 24 25 26 27 28 29 30

13. Please add below any comments that you feel are pertinent to a survey of the status of general mathematics in American colleges.

Name ....   School ....

APPENDIX C

LETTER B

(Second request sent to above colleges which failed to answer first request.)

A few months ago, I wrote for some information concerning college general mathematics as offered in your institution. I am sure that during the rush that comes at the close of school, the letter was misplaced and I apologize for sending it at such a time.

Many answers have been received, but to make the survey complete we need information which only you can give. Although all the information requested on the form is important to the survey, if there are certain questions that you prefer not to answer, draw a line through these items. But in any case, please fill in the enclosed form as completely as possible and mail in the self-addressed stamped envelope NOW before it is laid aside and forgotten.

I greatly appreciate your cooperation, and any comments concerning the reaction of your faculty and students to general mathematics will be most welcome.

APPENDIX D

QUESTIONNAIRE B

I. Please underline the names of the mathematics courses that you had in high school, and also underline the phrase that best describes your opinion of the course. Try to make this your opinion of the course and not the teacher.

General Mathematics
    a. Very Interesting   b. Fairly Interesting   c. Of Little Interest
Elementary Algebra
    a. Very Interesting   b. Fairly Interesting   c. Of Little Interest
Plane Geometry
    a. Very Interesting   b. Fairly Interesting   c. Of Little Interest
Solid Geometry
    a. Very Interesting   b. Fairly Interesting   c. Of Little Interest
Trigonometry
    a. Very Interesting   b. Fairly Interesting   c. Of Little Interest
Intermediate or
Advanced Algebra
    a. Very Interesting   b. Fairly Interesting   c. Of Little Interest
Business Arithmetic
    a. Very Interesting   b. Fairly Interesting   c. Of Little Interest

II. Please underline the phrase that best describes your opinion of *this* course.

    a. Very Interesting   b. Fairly Interesting   c. Of Little Interest

III. Please further indicate your opinion of *this* course by listing those parts you have enjoyed most and those parts that you have enjoyed least. In other words, just tell what you think of this course.